Evolution

*Its Nature, its Evidences and its
Relation to Religious Thought*

JOSEPH LeCONTE

CAMBRIDGE
UNIVERSITY PRESS

CAMBRIDGE UNIVERSITY PRESS

Cambridge, New York, Melbourne, Madrid, Cape Town, Singapore,
São Paolo, Delhi, Dubai, Tokyo

Published in the United States of America by Cambridge University Press, New York

www.cambridge.org
Information on this title: www.cambridge.org/9781108000673

© in this compilation Cambridge University Press 2009

This edition first published 1898
This digitally printed version 2009

ISBN 978-1-108-00067-3 Paperback

CAMBRIDGE LIBRARY COLLECTION

Books of enduring scholarly value

Religion

For centuries, scripture and theology were the focus of prodigious amounts of scholarship and publishing, dominated in the English-speaking world by the work of Protestant Christians. Enlightenment philosophy and science, anthropology, ethnology and the colonial experience all brought new perspectives, lively debates and heated controversies to the study of religion and its role in the world, many of which continue to this day. This series explores the editing and interpretation of religious texts, the history of religious ideas and institutions, and not least the encounter between religion and science.

Evolution

Joseph LeConte was the first geologist, natural historian and botanist to be appointed to the University of California in 1869. He founded the successful palaeontology program at Berkeley and acquired important collections of fossils. He also lectured and wrote on evolution, of which he was the leading American proponent. This book, first published in 1888 but revised and expanded in the second edition reissued here, is his attempt to reconcile his evolutionist convictions with his religious faith. Such a synthesis, he felt, was impeded by dogmatism on both sides, and he makes a case for 'a combining, reconciling and rational view'. He considers three questions: What is evolution? Is it true? and What then?, intending to address 'the intelligent general reader' without being superficial or unscientific. Concepts such as 'neo-Darwinism', 'materialism', and 'design' make their appearance in this wide-ranging book, whose concerns remain surprisingly topical today.

Cambridge University Press has long been a pioneer in the reissuing of out-of-print titles from its own backlist, producing digital reprints of books that are still sought after by scholars and students but could not be reprinted economically using traditional technology. The Cambridge Library Collection extends this activity to a wider range of books which are still of importance to researchers and professionals, either for the source material they contain, or as landmarks in the history of their academic discipline.

Drawing from the world-renowned collections in the Cambridge University Library, and guided by the advice of experts in each subject area, Cambridge University Press is using state-of-the-art scanning machines in its own Printing House to capture the content of each book selected for inclusion. The files are processed to give a consistently clear, crisp image, and the books finished to the high quality standard for which the Press is recognised around the world. The latest print-on-demand technology ensures that the books will remain available indefinitely, and that orders for single or multiple copies can quickly be supplied.

The Cambridge Library Collection will bring back to life books of enduring scholarly value (including out-of-copyright works originally issued by other publishers) across a wide range of disciplines in the humanities and social sciences and in science and technology.

EVOLUTION

ITS NATURE, ITS EVIDENCES, AND ITS RELATION TO RELIGIOUS THOUGHT

BY

JOSEPH LE CONTE

AUTHOR OF "RELIGION AND SCIENCE," ETC.
AND PROFESSOR OF GEOLOGY AND NATURAL HISTORY IN THE
UNIVERSITY OF CALIFORNIA

SECOND EDITION, REVISED

NEW YORK
D. APPLETON AND COMPANY
72 FIFTH AVENUE
1898

PREFACE TO THE SECOND EDITION.

THE three years which have elapsed since the publication of the first edition of this work have been years of great activity of thought on many of the subjects treated therein. Some changes and additions seemed therefore imperatively called for.

For example: There has sprung up recently among the foremost writers on evolution a warm discussion on the *factors* of evolution, their number and relative importance. I have therefore added a chapter (Chap. III, Part II) on this subject—not, indeed, to discuss it fully (for this would be impossible in the limits of a chapter), but to put the mind of the reader in position to understand it and to judge for himself.

Again: Every reader of the first edition must have remarked that there are many fundamental religious questions which I have not touched at all in Part III. I had avoided these because my own mind was not yet fully clear. I regarded what I then wrote as only a little leaven in a very large lump. I was willing to wait and let it work. In the mean time it has worked in my own mind, and I hope in the minds of others. I have therefore added two chapters to this part. In one I simply carry out to their logical consequences the doctrine of the Divine Immanency. This brings up the questions of *First and Second Causes;* of *General and Special Providence;* of the *Natural and the Supernatural;* of *Mind vs. Mechanics in Nature,* etc., and shows the necessary changes of view which are enforced by the theory of evolution.

In the other I take up very briefly " *The Relation of Evolution to the Doctrine of the Christ.*" In the discussion of this I restrain myself strictly within the limits of the subject as stated above.

The only other important changes are in Chapter IV, Part III, " *On the Relation of Man to Nature.*" As I regard this as the most important chapter in the whole book, I have endeavored still further to enforce my view of the origin of man's spirit, and especially to make it clearer by means of several additional illustrations.

JOSEPH LE CONTE.

BERKELEY, CAL., *July 1, 1891.*

PREFACE TO THE FIRST EDITION.

THE subject of the following work may be expressed in three questions: What is evolution? Is it true? What then? Surely, there are no questions of the day more burning than these. Much has been written on each of them, addressed to different classes of minds: some to the scientific, some to the popular, and some to the religious and theological; but nothing has yet appeared which covers the whole ground and connects the different parts together. Much, very much has been written, especially on the nature and the evidences of evolution, but the literature is so voluminous, much of it so fragmentary, and most of it so technical, that even very intelligent persons have still very vague ideas on the subject. I have attempted to give (1) a very con-cise account of what we mean by evolution, (2) an out-line of the evidences of its truth drawn from many differ-ent sources, and (3) its relation to fundamental religious beliefs. I have determined, above all, to make a book so small that it may be read through without much ex-pense of time and patience. But the subject is so large that in order to do so it was necessary to sacrifice all but what was most essential, and to forego all redundancy (the bane of so-called popular science) even at the risk of baldness and obscurity. Nevertheless, I hope that the first and second parts will be found not only interesting to the intelligent general reader, but even profitable to the special biologist. I have tried to make these parts as untechnical as possible, but I hope not on that account the less scientific. For I am among those who think that it is not necessary to be superficial in order to be popular—that science may be adapted to the intelligent popular mind without ceasing to be science.

The third part seems to me still more important just now. There is a deep and widespread belief in the popular mind, and even to some extent in the scientific mind, that there is something exceptional in the doctrine of evolution as regards its relation to religious thought and moral conduct. Other scientific theories have required only some modifications of religious conceptions, but this utterly destroys the possibility of all religious belief by demonstrating a pure materialism. Now this, I believe, is a complete misconception. Thinking men are fast coming to see this; some, indeed, have mistaken the change for a reaction against evolution. It is a reaction not against evolution, but only against its materialistic implication. Evolution is more and more firmly established every year. The tide of conviction is one which knows no ebb. Some clear statement, in brief space, of its true relation to religious thought seems, therefore, very important at this time.

BERKELEY, CAL., *May, 1887.*

CONTENTS.

PART I.

WHAT IS EVOLUTION?

CHAPTER I.

ITS SCOPE AND DEFINITION.

PAGE

A type of evolution—Development of an egg 3

Universality of evolution—Pervades all nature and concerns all departments of thought—One half of all science—Illustrated (1) by human body, (2) by solar system, (3) by society, (4) by earth, (5) by organic kingdom—The term evolution usually, but not rightly, confined to this last 3

Definition of evolution—*I. Progressive change*—Shown in the animal body, or the *Ontogenic* series—In the animal scale, or the *Taxonomic* series—In the geological, or *Phylogenic* series—The three series similar, though not identical 8

II. Change according to certain laws—Three laws of succession of organic forms 11

(a) *Law of differentiation*—Early forms are generalized; afterwards separated into specialized forms—Illustrated by fishes, by birds—Whole process of differentiation illustrated by growth and branching of a tree 11

(b) *Law of progress of the whole*—Mistake of confounding evolution with upward progress—How far true, and how far false—Illustrated by branching tree—Examples of this mistake in the popular mind—In the scientific mind 13

(c) *Law of cyclical movement*—Shown in geological history—Age of mollusks, fishes, reptiles, mammals, man—Illustrated again by

PAGE

a branching tree—Increasing complexity as well as height—
Illustrated by diagram 16
The above three laws are laws of evolution—Differentiation—Shown
in the development of an egg, the type of evolution . . . 19
Progress of the whole—Not progress of all parts—Shown in the development of an egg 22
Cyclical movement—Less fundamental than other two—Shown in
Ontogeny of body, of mind—Increasing complexity—Necessity
of continued advance—Otherwise deterioration—All these laws
shown in progress of society—Differentiation shown—Progress
of the whole but not of all parts shown—Cyclical movement
shown—In social evolution, however, there is another element,
viz., conscious voluntary progress—This kind of evolution contrasted with the other 22
III. Change by means of resident forces—This is the point of dispute—Sense in which we use term resident forces—Does not
touch question of origin of natural forces 27
The two views of the origin of organic forms briefly contrasted—As
to whether natural or supernatural—As to variability, definite
or indefinite—As to change from one species to another by
transmutation or substitution—As to universality of law of continuity 29

CHAPTER II.

THE RELATION OF LOUIS AGASSIZ TO THE THEORY OF EVOLUTION.

General misunderstanding on this subject—Necessary to give sketch
of history of the idea—Greeks, Lucretius, Swedenborg, and
Kant—First scientific presentation by Lamarck—General character of Lamarck's views—Failed, and rightly so—Next, Chambers's " Vestiges of Creation "—Its general character—Failed,
and rightly so—Some think this unfortunate—Why not so—
An obstacle must be removed and a basis laid 32
The obstacle removed—Old views in regard to forces—Correlation of
forces established—But vital force considered exception—Therefore living forms also supposed exception to mode of origin of
other forms—Then vital forces also correlated—Therefore, *a
priori* probable that living forms also correlated with other
forms as to mode of origin—Thus obstacle removed . . . 35

PAGE

The basis laid—Agassiz laid inductive *basis* of evolution, although he refused to build—He established the laws of evolution and perfected the method of comparison—Importance of method discussed—The method of *notation*—The method of experiment—The difficulty of applying these to life phenomena—Method of comparison shown—(1) In Taxonomic series—(2) In Ontogenic series—(3) In Phylogenic series—Cuvier the great worker by comparison in the Taxonomic series—Agassiz in the Ontogenic and Phylogenic—Agassiz also established the three laws of evolution given in previous chapter—Thus he laid foundation—Why he did not build—Supposed identity of evolution and materialism—The obstacle being removed and the basis laid, when evolution again brought forward it was universally accepted, because the world was prepared—Place of Agassiz and Darwin compared—Formal science *vs.* physical science—Illustrated by relation of Kepler to Newton—Relation of Agassiz to time cosmos similar to that of Kepler to space cosmos—So Darwin to Newton—Some reflections on the above—Gravitation is the law of space cosmos—Evolution of time cosmos—Of the divine spheral music gravitation is the chordal harmony and evolution the melody 37

PART II.

EVIDENCES OF THE TRUTH OF EVOLUTION.

CHAPTER I.

GENERAL EVIDENCES OF EVOLUTION AS A UNIVERSAL LAW.

Evolution is continuity, causal relation, gradual becoming—Increasing acceptance of this idea—First accepted for inorganic forms, mountains, continents and seas, rocks and soils, earth as a whole, heavenly bodies—Therefore acknowledged for all inorganics—Influence of geology in bringing about this change—Organic forms : acknowledged for individuals, true for classes, orders, families, genera—Races and varieties also formed gradually—Artificial species formed gradually—Examples of gradual changes in wild species—Hyatt's researches—Other examples —Summing up of general evidence—Sufficient ground for induc-

PAGE

tion—But evolution is not only inductively probable but certain, axiomatic—It is the law of causation applied to forms, and therefore a necessary truth 53

CHAPTER II.

SPECIAL PROOFS OF EVOLUTION.

Introductory.

Special proofs necessary—Evolution, though certain, is not yet accepted by the popular mind—Different departments from which proofs are drawn 67

Origin of new organic forms; the old view briefly stated—Necessary to give a brief statement of theories—Old view—Permanency of specific types—Supernatural origin of species—Centers of creation—Explanation of facts of geographical distribution—Of geological distribution—Modification of extreme view—Variability, but within limits—Illustrated 68

The new view briefly stated—Indefinite variability of organic forms—Effect of environment on rigid forms—On plastic forms—Taxonomic groups represent degrees of kinship 72

Factors of evolution—(1) Physical environment—(2) Use and disuse of organs—(3) Natural selection—(4) Sexual selection—(5) Physiological selection—Its necessity shown—Its operation explained—Compared with natural selection—Cause of variation unknown—Explanation of this is the next great step in the theory of evolution 73

CHAPTER III.

THE GRADES OF THE FACTORS OF EVOLUTION AND THE ORDER OF THEIR APPEARANCE.

Factors of evolution restated; their grades and the order of their introduction shown—Lamarckian factors, first in order because they precede sexual reproduction—(1) Environment—(2) Use and disuse—With sexual reproduction selective factors introduced—(3) Natural selection—(4) Physiological selection—(5) Sexual selection—With man was introduced (6) the rational factor—In this process two striking stages—viz., the introduction of sex and the introduction of reason—Effect of each to hasten the steps of evolution—The last by far the greater change . . 81

CONTENTS

PAGE

Contrast between organic and human evolution—(1) The meaning of
term *fittest* in each—(2) Destiny of the weak and helpless in
each—(3) The nature of evolutionary transformation in each
—(4) The law of strait and narrow way applied in each—(5)
Human evolution is a different kind and on a higher plane . 88
Application to some questions of the day.
I. Neo-Darwinists, their position explained—Reasons for dissenting
—(*a*) Lamarckian factors preceded all others—(*b*) Though now
subordinate, they still underlie and condition all other factors—
(*c*) Shown by comparison of phylogeny with ontogeny . . 92
II. Human progress not identical with organic evolution—Mistake
of the materialists—But neither is it wholly different, as some
suppose 96
III. Neo-Darwinism is fatal to hopes of human progress—Reason
may use freely Lamarckian factors, but can not use natural
selection in the same way as Nature does 97

CHAPTER IV.

SPECIAL PROOFS FROM THE GENERAL LAWS OF ANIMAL STRUCTURE, OR
COMPARISON IN THE TAXONOMIC SERIES.

General Principles.

Analogy and homology—Defined and illustrated by examples—Wings
and limbs—Lungs, gradual formation of, traced in the Taxonomic series—Traced in the Ontogenic series—Examples of
homology in plants : tuber, cactus-leaf, acacia-leaf—Definitions
repeated and further explained—Common origin is the only explanation of homology 99
Primary divisions of the animal kingdom—True ground of such divisions is ability to trace homology—We take examples only from
vertebrata and articulata—Compare to styles of architecture—
To machines—To branching stem 107

CHAPTER V.

PROOFS FROM HOMOLOGIES OF THE VERTEBRATE SKELETON.

Common general plan—In several respects—Strongly suggestive
of common origin—Details of structure demonstrative of the
same 111

PAGE

Special homology of vertebrate limbs 113
Fore-limbs—Comparison of fore-limbs of mammals, birds, reptiles,
and fishes, part for part—Gradual changes in collar-bone and
coracoid—In position of elbow—In bones of forearm—In position of wrist—In the tread—The term manus—Number of toes
—Modifications for flight in various animals—For swimming in
whales and fishes 113
Hind-limbs—Comparison of hind-limbs of several mammals—Position of knee—Of heel—Plantigrade and digitigrade—Degrees
of the latter—Number of toes—General law in regard to number of similar parts—Order of toe-dropping in artiodactyles—
In perissodactyles 121
Genesis of the horse—Changes in foot-structure—Same true of other
parts of skeleton—Only natural explanation is derivation—Nature compared with man in mode of working—Angels—Griffins
—Centaurs—Muscular and nervous systems—Visceral organs . 126

CHAPTER VI.

HOMOLOGIES OF THE ARTICULATE SKELETON.

Illustrations from this type—Plan of structure entirely different—
General plan of structure explained and modifications shown—
Shrimp—Modification of segments and of appendages for various purposes: swimming, walking, eating, sense—Illustrated
by other crustaceans—By myriapods—By marine worms—
Crabs—Embryonic development of crabs—insects—Modification of segments and appendages—Mouth parts of insects . 132
Illustration of the law of differentiation—Cells—Segments—Individuals—Homologies in other departments of animals, but these
are less familiar—Between primary groups, homology untrace-
able in adult forms—But these also probably connected by common origin—Different views as to origin of vertebrates . . 144

CHAPTER VII.

PROOFS FROM EMBRYOLOGY, OR COMPARISON IN THE ONTOGENIC SERIES.

Resemblance of the three series—Frog, in Ontogeny passes through
main stages of Taxonomy and Phylogeny—Resemblance only
general—Many steps dropped out in Ontogeny 148

PAGE

(1) *Ontogeny of tailless amphibians*—The frog : fish stage, perenni-branch stage, caducibranch stage, aneural stage—Same stages in Phylogeny 150

(2) *Aortic arches*—Those of lizard described—Origin from gill-arches of fish—Change from one to the other in Ontogeny of a frog —Same changes in Phylogeny of lizard—Embryonic condition of mammalian heart and vessels—Gradual change and final condition in birds—In mammals—Gradual decrease in number of aortic arches as we go up the vertebrate scale—Cogency of the argument from aortic arches 151

(3) *Vertebrate brain*—Fish brain—Brain of reptiles, birds, mammals, man compared—Human brain passes through similar stages—Changes in complexity of structure in Taxonomy—Same changes in Ontogeny of mammals—Same in Phylogeny of reptiles, birds, mammals 162

Cephalization—Explanation of, in body, in mind 171

(4) *Fish-tails*—Homocercal and heterocercal—Vertebrated and non-vertebrated—Order of change in Ontogeny—Same in Phylogeny —Similar changes in birds' tails in Ontogeny and Phylogeny—In other tailless animals—Examples from articulates, insects, crustaceans, etc. 172

Illustration of the differentiation of the whole animal kingdom—Development of eggs of all kinds of animals—This a type of changes in Phylogeny—Why Ontogeny repeats Phylogeny—Law of acceleration 176

Proofs from rudimentary and useless organs—Examples from whale : Teeth—Limbs—Hair—Olfactive organs—Examples from man : muscles, cæcal appendage—Significance of useless organs . 179

CHAPTER VIII.

PROOFS FROM GEOGRAPHICAL DISTRIBUTION OF ORGANISMS.

Geographical faunas and floras—Conditions which limit . . . 183

Temperature-regions—Illustrated by plants—In latitude and in elevation—Same in animal species 184

More perfect definition of regions—Range of different Taxonomic groups—Gradual shadings on borders of range—Shadings out of individuals in number and vigor, but not in specific character —As if centers of origin—Effect of east and west barriers—

PAGE

Temperature regions repeated south of the equator, but not
species—As if centers of origin 186

Continental faunas and floras—Temperature zones continuous, but
not species—Reason: ocean barriers—As if centers of origin—
Polar regions: one. Why—Temperate zone—Different species
in different continents—Species of United States and of Europe
almost wholly different—As if origin local—Exceptions—(1)
Introduced species—(2) Hardy or else wide-migrating species—
(3) Alpine species—Tropical zone of two continents still more
different—Same true of south temperate zone 188

Subdivisions of continental faunas and floras—Illustrated by fauna
and flora of United States 191

Special Cases—Australia—Madagascar—Galapagos—River mussels. 192

Marine species—Same principles applicable—Therefore organic
forms grouped in regions, sub-regions, provinces, etc.—Primary
regions according to Wallace—According to Allen . . . 192

Theory of the origin of geographical diversity—Specific centers of
creation—Objections to. The element of time left out—Progressive change in unlimited time, or evolution the only rational
explanation—This connects with geographical changes in geological times, especially the Glacial epoch—Geographical diversity in other times 193

Most probable view of the general process—Last great period of
change was the Glacial epoch—This, therefore, is the key to
geographical distribution—Condition of things during the Glacial epoch—In America—Changes in temperature—In physical geography and in species—In Europe—Application of principles 196

(1) *Australia*—Characteristics of its fauna—Explanation of—Isolation very early—Position of marsupials and monotremes in the
Taxonomic scale—Australia isolated before the Tertiary—Effect
of competition on evolution 200

(2) *Africa*—African region defined—Two groups of its mammals,
indigenes, and invaders—Effect of the invasion . . . 204

(3) *Madagascar*—Characteristics of its fauna—Relation to African
indigenes—Separated before the invasion—Significance of its
lemurs 205

(4) *Island life*—Two kinds of islands—Defined and illustrated by
examples—(a) *Continental islands*—General character of fauna

CONTENTS.

PAGE

—Illustrated by Madagascar, New Zealand, British Islands, coast-islands of California—Characteristics of the faunas of these explained—(b) *Oceanic Islands*—Defined—Characteristics of faunas and their origin 207

(5) *Alpine species*—Characteristics of and their origin explained— Migrations of Arctic species during Glacial times, and their isolation on mountains 215

Objection—Mode of change of species on borders of ranges—Examples—Sweet-gum—Sequoia 217

Answer—Distribution of these forms in time, and their migrations —They are remnants—Intermediate forms are extinct . . 219

CHAPTER IX.

PROOFS FROM VARIATION OF ORGANIC FORMS, ARTIFICIAL AND NATURAL.

Limitation of the use of experiment in morphology—Unconscious experiments in breeding, and their results—Principles involved— Inheritance, immediate and ancestral—Effect of true breeding long continued—Method of selection illustrated by diagram— Formation of a race—Process the same in nature—Show selective effect of physical environment—Of organic environment— f migrations—Of unlimited time—Other factors of change, and their effects shown in nature and in domestication—Differences between artificial and natural species 222

First difference, reversion—The tendency to reversion described— The reason explained—Illustrated by the case of the pointer . 229

Second difference, intermediate forms—Reason is, these are eliminated in nature 232

Third difference, cross-fertility—Natural species are usually crosssterile—Degrees of cross-sterility—Two bases of species, morphological and physiological—Two kinds of isolation, sexual repugnance and cross-sterility—Latter most essential—Illustrated by plants and hermaphrodite animals—Former only higher animals—Natural laws interfered with by domestication —Illustrated by plants and animals 232

Law of cross-breeding—Effect of close breeding—Of crossing varieties to a limit—The law investigated—Reproduction in lowest organisms—Fission—Gemmation—Internal gemmation—Sex introduced—Effect of, is funding of differences in offspring and

PAGE

tendency to variation—Sexual and non-sexual reproduction compared—Separation of sex elements—Of sex-individuals—Introduction of sex-attraction—Funding of greater differences in offspring—Crossing of varieties—Diagram illustrating effect in vigor—Effect also in plasticity—Application of these principles —Necessity of sexual isolation to produce species—Origin of cross-sterility and thus of species by Dr. Romanes's idea—Why artificial varieties are cross-fertile—Geographical species sometimes cross-fertile—Application of principles—Absence of intermediate links in natural species explained—Under what conditions such are found—Further explanation of this point—Illustrated by a growing tree 236

Objection answered—Intermediate links ought to be found fossil—Answer (1) Imperfection of record. (2) The term species indefinite. (3) Transitions between all other taxonomic groups abundant. (4) Between species, also, both living and fossil—Of fossil, Planorbis of Steinheim—Other examples—(5) Why transition-forms are rare—Answer—Changes in every department of nature are paroxysmal—Illustrated—So the steps of evolution paroxysmal—Critical periods in evolution—Causes of rapid advance—Apparent discontinuity between species—(1) changes paroxysmal—(2) Brooks's idea—Male sex is the progressive element—Illustrated by society—Effect of prosperous times—Mrs. Treat's experiments—Hard times produce excess of males, and therefore tend to diversity—Summary . . . 248

Objection—Egyptian drawings and mummy plants, show no change —Answer (1) Time too short. (2) We are now in time of slow change. (3) All species don't change, most become extinct. (4) Evolution is probably slower now than formerly—Reasons for so thinking—Organic evolution approaching completion—Other supposed objections 265

Origin of beauty—Explanation of, in higher animals—In flowering plants—But in many cases we can't explain 269

Incipient organs—Difficulty of explaining—But these are not objections to the *fact* of evolution, but only to the sufficiency of the present *theories* of evolution. Therefore, all discussion concerns special theories. The fact of evolution is certain . . . 270

PART III.

THE RELATION OF EVOLUTION TO RELIGIOUS THOUGHT.

CHAPTER I.

INTRODUCTORY.

PAGE

Evolution if true affects every department of thought—What will be its effect on religious beliefs ?—Objection that truth-seeker has nothing to do with effects—Answered 275

Relation of the true and the good 277

Relation of philosophy to life—The three necessary elements of a rational philosophy—Application to the case in hand—And the subject of Part III justified—Exaggerated fears—Different forms of the conflict of science and religion—(1) Heliocentric theory —First effect and final result—(2) Law of gravitation—Effect and result—(3) Antiquity of the earth and cosmos—Effect and result—(4) Antiquity of man—(5) Evolution 277

CHAPTER II.

THE RELATION OF EVOLUTION TO MATERIALISM.

Supposed identity—Tendency of the age—Evolution does not differ in this regard from other laws of Nature—Absurdity of identification illustrated in many ways—(1) Effect of discovery of process of making—(2) Effect of new form of old truth—(3) Manner in which vexed questions are settled and rational philosophy found—Illustrated—A true philosophy is a reconciliation of partial views—Three possible views of origin of individuals and of species ; two one-sided and partial, and the third combining, reconciling, and therefore rational—The only bar to speedy reconciliation is dogmatism—Theological and scientific —The appropriate rebuke for each—Thereforee volution does not differ from other laws in regard to its relation to materialism— Nevertheless, great changes in our traditional beliefs impending —Main changes are notions concerning God, Nature, and man, in their relations to one another 284

2

CONTENTS.

CHAPTER III.

THE RELATION OF GOD TO NATURE.

PAGE

The issue in regard to this relation stated—The growth of the issue
described—The old view of direct relation—The effect of science
and the resulting view—The compromise—Destroyed by evolution—The issue forced—The alternative view—Immanence of
Deity—This view explained—Objection of idealism—Answered
—It is not subjective idealism—Objection of pantheism—Answer deferred—Objection that the view is incompatible with
practical life—Answered 297

CHAPTER IV.

THE RELATION OF MAN TO NATURE.

The two extreme views in this regard—They are views from different
points, psychical and material—The latter very productive in
modern times—But many fear the final effect—Reconciliation is
possible—Scientific materialism has two branches—*Physiological branch* explained—Conclusion—Answer—Relation of psychic
to brain changes is inscrutable—The mystery illustrated—Outside and inside view—Different from other phenomena in this
regard 304
Evolution branch—Close relation of man to animals—Therefore must
extend immortal spirit to animals—to plants—to all existence,
and thus identify immortality with conservation of force—Embryonic series—Where did spirit enter ?—Evolution series—
Where did spirit enter ?—Answer—Derived from Nature—The
true view of origin stated—Show that it is not in discord with
other phenomena of evolution—The five planes of matter and of
force—The change from one to another not gradual now nor in
the evolution of natural forces—Consecutive births into higher
forms—Every step of these changes taking place now—Relation of these facts to immortality—The process briefly stated—
Omnipresent divine energy individuated to separate entity in
man—Anima of animals is spirit in embryo—Came to birth in
man—Illustrated in other ways—(1) By more or less completed
water-drop—(2) By submergence and emergence—(3) By planet
birth—(4) By physical birth—(5) By grades of organic indi-

PAGE
viduality—(6) By the body as an instrument of communication
between two worlds—Self-consciousness the sign of spirit-individuality—Any animal conscious of self would be immortal—
Similar changes in passing from animals to man in all other departments of psychic activity—Objection that other changes of
energy not permanent; answered—Our view of origin compared
with alternative views—Plato's view—Orthodox view .   . 311
Some general conclusions—(1) Two series of changes, brain-changes
and mind-changes—The initiative in animals—In man—(2)
Justification of term "*vital principle*"—Becomes entity in man
—(3) This view is a complete reconciliation of realism and
nominalism—(4) No meaning in Nature without spirit—And no
meaning in geological history without derivative origin of spirit
—Material evolution finds its goal in man, psychic evolution in
the divine man 327

CHAPTER V.

THE RELATION OF GOD TO MAN.

Question of revelation—Difficulty of the subject—Operation of divine
spirit on spirit of man more direct than on Nature—This is revelation—This is no violation of law, but operation by higher law
—Term supernatural is relative—Illustrated—There is but one
kind of revelation, and this to all men in different degrees—
Always imperfect, and therefore must be tried by reason .   . 331

CHAPTER VI.

THE OBJECTION, THAT THE ABOVE VIEW IMPLIES PANTHEISM, ANSWERED.

The objection stated and the general answer—In deepest questions
single lines of thought lead to extreme views—Must follow other
lines—These lead to personality 335
(1) Exact character of relation of God and of necessary law to man's
freedom is inscrutable 338
(2) On the inside of brain-changes we find personality—So on the inside of natural phenomena must also be person—In either case
science studies the outside only—In Nature all is mechanics on
the outside, but all is mind on the inside—Thought behind brain-
changes compels belief in same behind natural phenomena—
Law of infinite expansion—Illustrated by ideas of Space and

PAGE

Time—So also with idea of self—Infinite person inconceivable,
but contrary is more inconceivable—Illustrated by ideas of
Space and Time 338
(3) Idea of *Causation* and of *Force*—Derived from *within*—Steps of
the evolution of this idea—Final result is one infinite personal
will—Expansion of idea of causal nexus between phenomena to
the idea of one infinite cause 342
(4) Idea of *design* also originates *within*—Ineradicable, but changes
form—Expands to infinity—Same change produced by science
in all our notions concerning God—Same in our sense of *mystery*—Same in our notions concerning *creation*—Same in our
conceptions of *design*—Thus, self-consciousness behind brain-
changes compels belief in God behind Nature—The closeness
of connection in the one case necessitates closeness of connection in the other—Every material change in Nature caused by
a mental change behind Nature 345

CHAPTER VII.

SOME LOGICAL CONSEQUENCES OF THE DOCTRINE OF THE DIVINE IMMANENCY.

Religious thought subject to the law of evolution; three main stages 351
I. *Conception of God*—The three stages shown—(1) Anthropomorphism—(2) Absentee landlordism—(3) Immanence . . 351
II. *Question of First and Second Causes*—The three stages shown
here—(1) All is First Cause but man-like—(2) Distinction of
first and second causes introduced—(3) Identification of these 354
III. *General and Special Providence*—The same three stages shown
and the same outcome—viz., identification 355
IV. *Natural and the supernatural*—The same stages and the same
final identification—Question of miracles 355
V. *Question of design or mind in Nature*—The same three stages
and the same solution shown here—Confusion in the minds of
modern writers 357
VI. *Question of mode of creation*—Old and new views contrasted . 358

CHAPTER VIII.

RELATION OF EVOLUTION TO THE IDEA OF THE CHRIST.

Comparison of organic with human evolution—The idea of the first
is *man*, of the second is *the Christ*—Definition of the Christ as

ideal man—The Christ ought to differ from us in a superhuman
way—Shown by several illustrations—The Christ, as ideal man,
a true object of rational worship—The ideal of organic evolution comes *at the end*—Ideal of human evolution must come *in
the course*—Objection that there are many partial ideals answered—Relative *vs.* absolute moral ideal   360

CHAPTER IX.

THE RELATION OF EVOLUTION TO THE PROBLEM OF EVIL.

The difficulty of the problem—The light on it by evolution—Evil
must be based on the constitution of Nature and therefore universal—Some of its forms   365
(1) *Physical evil in animal kingdom*—Condition of organic evolution is struggle with an apparently inimical environment—In
its course it seems evil—Looking back from the end it is
good   365
(2) *Physical evil in relation to man*—Necessary condition of social
evolution is also struggle with a seeming evil environment—But
looking back from the end this evil is also seen to be good—
Without it man would never have emerged from animality   . 366
(3) *Organic evil—Disease*—This also is the necessary condition of
acquisition of knowledge of organic Nature—In the course of
evolution it seems evil, but from the end it is seen to be good—
In the physical world, laws of Nature are beneficent in their
general operation, and only evil in their specific operation through
our ignorance 367
(4) *Moral evil—Moral disease*—Difference between this and other
forms of evil—Can this also be transmuted into good ?—This is
only the highest form of evil, and therefore subject to the same
laws of evolution—Here also elevation comes only through
knowledge and power, and these only through struggle with apparent evil—In course it seems evil, looking back from end it
is seen to be good to the race—In all, therefore, the individual
is sacrificed to the race, but impossible here—A way of escape
found in the nature of a moral being—In this case not only
final victory for the race, but also within the power of the individual—In this case success is in proportion to honest effort
in right spirit—Roots of evil in the necessary law of evolution

PAGE

-It is the necessary condition of all progress—Without it a
moral being is impossible—From philosophic point of view
things are not good and evil, but only higher and lower—All
things good in their places—Evil is discord—Good is due relation—Action and reaction of higher and lower is the necessary
ndition of true virtue 369

PART I.

WHAT IS EVOLUTION?

CHAPTER I.

ITS SCOPE AND DEFINITION.

A Type of Evolution.—Every one is familiar with the main facts connected with the development of an egg. We all know that it begins as a microscopic germ-cell, then grows into an egg, then organizes into a chick, and finally grows into a cock ; and that the whole process follows some general, well-recognized law. Now, this process is evolution. It is more—it is *the* type of all evolution. It is that from which we get our idea of evolution, and without which there would be no such word. Whenever and wherever we find a process of change more or less resembling this, and following laws similar to those determining the development of an egg, we call it evolution.

Universality of Evolution.—Evolution as a *process* is not confined to one thing, the egg, nor as a doctrine is it confined to one department of science—biology. The process pervades the whole universe, and the doctrine concerns alike every department of science—yea, every department of human thought. It is literally one half of all science. Therefore, its truth or falseness, its acceptance or rejection, is no trifling matter, affecting only one small corner of the thought-realm. On the contrary, it affects profoundly the foundations of philosophy, and therefore the whole domain of thought. It determines the whole attitude of the mind toward Nature and God.

I have said evolution constitutes one half of all science. This may seem to some a startling proposition. I stop to make it good.

Every system of correlated parts may be studied from two points of view, which give rise to two departments of science, one of which—and the greater and more complex—is evolution. The one concerns changes within the system by action and reaction between the parts, producing equilibrium and stability; the other concerns the progressive movement of the system, as a whole, to higher and higher conditions — the movement of the point of equilibrium itself, by constant slight disturbance and readjustment of parts on a higher plane, with more complex inter-relations. The one concerns the laws of sustentation of the system, the other the laws of evolution. The one concerns things as they are, the other the process by which they become so. Now, Nature as a whole is such a system of correlated parts. Every department and sub-department of Nature, whether it be the solar system or the earth, or the organic kingdom, or human society, or the human body, is such a system of correlated parts, and is therefore subject to evolution. We can best make this thought clear by examples:

1. Take, then, the *human body*. This complex and beautiful system of correlated and nicely-adjusted parts may be studied in a state of maturity and equilibrium, in which all the organs and functions by action and reaction co-operate to produce perfect stability, health, and physical happiness. This study is physiology. Or else the same may be studied in a state of progressive change. Now, we perceive that the stability is never perfect—the point of equilibrium is ever moving. By the ever-changing number and relative power of the co-operating parts the equilibrium is ever being disturbed, only to be readjusted on a higher plane, with still more beautiful and complex inter-relations. This is growth, development, evolution. Its study is called embryology.

2. Take another example—*the solar system*. We may study sun, planets, and satellites in their mutual actions and reactions, co-operating to produce perfect equilibrium, stability, beautiful order, and musical harmony. This is the ideal of physical astronomy as embodied in Laplace's "Mécanique Céleste." Or we may study the same in its origin and progressive change. Now, we perceive that equilibrium and stability are never absolutely perfect, but, on the contrary, there is continual disturbance with readjustment on a higher plane—continual introduction of infinitesimal discord, only to enhance the grandeur and complexity of the harmonic relations. This is the nebular hypothesis—the theory of the development of the solar system. It is cosmogony; it is evolution. 3. Again : *society* may be studied in the mutual

play of all its social functions so adjusted as to produce
social equilibrium, happiness, prosperity, and good gov-
ernment. This is social statics. But equilibrium and
stability are never perfect. Permanent social equilibrium would be social stagnation and decay. Therefore, we
must study society also in its onward movement—the
equilibrium ever disturbed, only to be readjusted on a
higher plane with more and more complexly inter-related
parts. This is dynamics—social progress. It is evolution. 4. Again: the *earth*, as a whole, may be studied
in its present forms, and the mutual action of all its parts
—lands and seas, mountains and valleys, rivers, gulfs,
and bays, currents of air and ocean—and the manner in
which all these, by action and reaction, co-operate to produce climates and physical conditions such as we now
find them. This is physical geography. Or, we may
study the earth in its gradual progress toward its present condition—the changes which have taken place in
all these parts, and consequent changes in climate ; in a
word, the gradual process of becoming what it now is.
This is physical geology—it is evolution. 5. Lastly, we
may study the whole *organic kingdom* in its entirety
as we now find it—the mutual relation of different
classes, orders, genera, and species to each other and to
external conditions, and the action and reaction of these
in the struggle for life—the geographical distribution of
species and their relation to climate and other physical
conditions, the whole constituting a complexly adjusted
and permanent equilibrium. This is a science of great

importance, but one not yet distinctly conceived, much less named.* Or, we may study the same in its gradual progressive approach, throughout all geological times, toward the present condition of things, by continual changes in the parts, and therefore disturbance of equilibrium and readjustment on a higher plane with more complex inter-relations. This is development of the organic kingdom. In the popular mind it is, *par excellence*, evolution.

We might multiply examples without limit. There are the same two points of view on all subjects. As already said, in the one we are concerned with things as they are ; in the other, with the process by which they became so. This "law of becoming" in all things —this universal law of progressive inter-connected change —may be called the law of continuity. We all recognize the universal relation of things, gravitative or other, in space. This asserts the universal causal relation of things in *time*. This is the universal law of evolution.

But it has so happened that in the popular mind the term evolution is mostly confined to the development of the organic kingdom, or the law of continuity as applied to this department of Nature. The reason of this is that this department was the last to acknowledge the supremacy of this law; this is the domain in which the advocates of supernaturalism in the realm of Nature had

---

* The term *Chorology*, used by Haeckel, nearly covers the ground.

made their last stand. But it is wholly unphilosophical
thus to limit the term. If there be any evolution, *par
excellence*, it is evolution of the individual or embryonic
development. This is the clearest, the most familiar,
and most easily understood, and therefore the type of
evolution. We first take our idea of evolution from this
form, and then extend it to other forms of continuous
change following a similar law. But, since the popular
mind limits the term to development of the organic
kingdom, and since, moreover, this is now the battle-
ground between the advocates of continuity and discontinuity—of naturalism and supernaturalism in the *realm
of Nature*—what we shall say will have reference chiefly
to this department, though we shall illustrate freely by
reference to other forms of evolution.

DEFINITION OF EVOLUTION.

Evolution is (1) continuous *progressive change*, (2)
according to certain laws, (3) and by means of *resident
forces*. It may doubtless be defined in other and perhaps better terms, but this suits our purposes best.
Embryonic development is the type of evolution. It
will be admitted that this definition is completely realized in this process. The change here is certainly continuously progressive ; it is according to certain well-
ascertained laws ; it is by forces (vital forces) resident
in the egg itself. Is, then, the process of change in
the organic kingdom throughout geologic times like
this ? Does it correspond to the definition given

above ? Does IT also deserve the name of evolution ? We shall see.

I. **Progressive Change.**—Every individual animal body —say man's—has become what it now is by a gradual process. Commencing as a microscopic spherule of living but apparently unorganized protoplasm, it gradually added cell to cell, tissue to tissue, organ to organ, and function to function; thus becoming more and more complex in the mutual action of its correlated parts, as it passed successively through the stages of germ, egg, embryo, and infant, to maturity. This ascending series of genetically connected stages is called the embryonic or *Ontogenic* series.*

There is another series the terms of which are coexistent, and which, therefore, is not in any sense a genetic or development series, but which it is important to mention, because to some degree similar to and illustrative of the last. Commencing with the lowest unicelled microscopic organisms, and passing up to the animal scale, *as it now exists*, we find a series of forms similar, though not identical, with the last. Here, again, we find cell added to cell, tissue to tissue, organ to organ, and function to function, the animal body becoming more and more complex in structure, in the mutual action of its correlated parts, and the mutual action with the environment, until we reach the highest complexity of structure and of internal and external relations only in the highest

---

* *Ontos-gennao* (individual-making, or genesis of the individual).

animals. This ascending series may be called the natural history series ; or, the classification or *Taxonomic series.** The terms of this series are, of course, not genetically connected ; at least, not directly so connected. In what way they are connected, and how the series comes to be similar to the last, we shall see by-and-by.

Finally, there is still a third series, the grandest and most fundamental of all, but only recently recognized, and therefore still imperfectly known. Commencing with the earliest organisms, the very dawn of life, in the very lowest rocks, and passing onward and upward through Eozoic, Palæozoic, Mesozoic, Cenozoic, to the Psychozoic or present time, we again find first the lowest forms, and then successively forms more and more complex in structure, in the interaction of correlated parts and in interaction with the environment, until we reach the most complex internal and external relations, and therefore the highest structure only in the present time.†
This series we will call the geological or *phylogenic* series.‡ According to the evolution theory, the terms of *this* series also are genetically connected. It is, therefore, an evolution series. Furthermore, it is the most fundamental of the three series, because it is the *cause* of the other two. The Ontogenic series is like it because it is a brief recapitulation, through heredity, as it were from memory, of its main points. The Taxonomic series

---

* *Taxis, nomos* (relating to science of arrangement).
† This statement is general ; it will be modified hereafter.
‡ *Phule-gennao* (kind-making) ; genesis of the race.

is like it because the *rate* of advance along different lines was different in every degree, and therefore every stage of the advance is still represented in a general way among existing forms. Some of these points will be explained more fully in future chapters, in connection with the evidences of the truth of evolution.

It will be admitted, then, that we find *progressive change* in organic forms throughout geological times. This is the first point in the definition of evolution.

II. **Change according to Certain Laws.**—We have shown continuously progressive change in organic forms during the whole geologic history of the earth, similar in a general way to that observed in embryonic development. We wish now to show that the *laws of change* are similar in the two cases. What, then, are the laws of succession of organic forms in geologic times ? I have been accustomed to formulate them thus : *a.* The law of differentiation ; *b.* The law of progress of the whole ; *c.* The law of cyclical movement.* We will take up these and explain them successively, and then, afterward, show that they are also the laws of embryonic development, and therefore the laws of evolution.

a. **Law of Differentiation.**—It is a most significant fact, to which attention was first strongly directed by Louis Agassiz, that the earliest representatives of any group, whether class, order, or family, were not what we

---

* This formulation of the laws of organic succession was given by me in 1860, before I knew anything of either Darwin's or Spencer's evolution. They were my own mode of formulating Agassiz's views.

3

would now call typical representatives of that group;
but, on the contrary, they were, in a wonderful degree,
connecting links ; that is, that along with their distinctive classic, ordinal, or family characters they possessed
also other characters which connected them closely with
other classes, orders, or families, now widely distinct,
without connecting links or intermediate forms. For
example : The earliest vertebrates were fishes, but not
typical fishes. On the contrary, they were fishes so
closely connected by many characters with amphibian
reptiles, that we hardly know whether to call some of
them reptilian fishes, or fish-like reptiles. From these,
as from a common vertebrate stem, were afterward separated, by slow changes from generation to generation, in
two directions, the typical fishes and the true reptiles.
So, also, to take another example, the first birds were far
different from typical birds as we now know them. They
were, on contrary, birds so reptilian in character, that
there is still some doubt whether bird-characters or reptilian characters predominate in the mixture, and therefore whether they ought to be called reptilian birds or
bird-like reptiles. From this common stem, the more
specialized modern reptiles branched off in one direction
and typical birds in another, and intermediate forms became extinct ; until *now*, the two classes stand widely
apart, without apparent genetic connection. This subject will be more fully treated hereafter, and other examples given. These two will be sufficient now to make
the idea clear.

Such early forms combining the characters of two or more groups, now widely separated, were called by Agassiz *connecting* types, *combining* types, *synthetic* types, and sometimes *prophetic* types ; by Dana, *comprehensive* types ; and by Huxley, *generalized* types. They are most usually known now as *generalized* types, and their widely-separated outcomes *specialized* types. Thus, in general, we may say that the widely-separated groups of the present day, when traced back in geological times, approach one another more and more until they finally unite to form common stems, and these in their turn unite to form a common trunk. From such a common trunk, by successive branching and rebranching, each branch taking a different direction, and all growing wider and wider apart (differentiation), have been gradually generated all the diversified forms which we see at the present day. The last leafy ramifications—flower-bearing and fruit-bearing—of this tree of life, are the fauna and flora of the present epoch. The law might be called a law of ramification, of specialization of the parts, and diversification of the whole.

b. **Law of Progress of the Whole.**—Many imagine that progress is the one law of evolution ; in fact, that evolution and progress are coextensive and convertible terms. They imagine that in evolution the movement must be upward and onward in all parts ; that degeneration is the opposite of evolution. This is far from the truth. There is, doubtless, in evolution, progress to higher and higher planes ; but not along every line, nor in every part ; for this would be contrary to the law of differentiation. It is only progress of the whole organic kingdom in its entirety. We can best make this clear by an illustration. A growing tree branches and again branches *in all directions*, some branches going upward some sidewise, and some downward—anywhere, everywhere, for light and air ; but the whole tree grows ever taller in its higher branches, larger in the circumference of its outstretching arms, and more diversified in structure. Even so the tree of life, by the law of differentiation, branches and rebranches continually in all directions—some branches going upward to higher planes (progress), some pushing horizontally, neither rising nor sinking, but only going farther from the generalized origin (specialization) ; some going downward (degeneration), anywhere, everywhere, for an unoccupied place in the economy of Nature, but the whole tree grows ever higher in its highest parts, grander in its proportions, and more complexly diversified in its structure.

It may be well to pause here a moment to show how this mistaken identification of evolution with progress alone, without modification by the more fundamental laws of differentiation, has given rise to misconceptions in the popular and even in the scientific mind. The biologist is continually met with the question, "Do you mean to say that any one of the invertebrates, such, for instance, as a spider, may eventually, in the course of successive generations, become a vertebrate, or that a dog

or a monkey is on the highway to become a man?" By
no means. There is but one straight and narrow way to
the highest in evolution as in all else, and few there be
that have found it—in fact, probably two or three only
at every step. The animals mentioned above have diverged from that way. In their ancestral history, they
have missed the golden opportunity, if they ever had
it. It is easy to go on in the way they have chosen,
but impossible to get back on the ascending trunk-
line. To compare again with the growing tree, only
one straight trunk-line leads upward to the terminal
bud. A branch once separated must grow its own way,
if it grow at all.

Of the same nature is the mistake of some extreme
evolutionists, such as Dr. Bastian and Professor Haeckel,
and of nearly all anti-evolutionists, viz., that of imagining that the truth of evolution and that of spontaneous
generation must stand or fall together. On the contrary, *if* life did *once* arise spontaneously from any lower
forces, physical or chemical, by natural process, *the conditions necessary for so extraordinary a change could
hardly be expected to occur but once in the history of the
earth.* They are, therefore, *now,* not only unreproduci-
ble, but unimaginable. Such golden opportunities do
not recur. Evolution goes only onward. Therefore, the
impossibility of the derivation of life from non-life *now,*
is no more an argument against such a derivation *once,*
than is the hopelessness of a worm ever becoming a vertebrate *now,* an argument against the derivative origin

of vertebrates. Doubtless if life were now extinguished
from the face of the earth, it could not again be rekin-
dled by any natural process known to us ; but the same is
probably true of every step of evolution. If any class—
for example, mammals—were now destroyed, it could not
be re-formed from any other class now living. It would
be necessary to go back to the time and conditions of the
separation of this class from the reptilian stem. Therefore, the falseness of the doctrine of abiogenesis,* so far
from being any argument against evolution, is exactly
what a true conception of evolution and knowledge of its
laws would lead us to expect.

c. **Law of Cyclical Movement.**—The movement of evolution has ever been onward and upward, it is true, but
not at uniform rate in the whole, and especially in the
parts. On the contrary, it has plainly moved in successive cycles. The tide of evolution rose ever higher and
higher, without ebb, but it nevertheless came in successive waves, each higher than the preceding and overborne
by the succeeding. These successive cycles are the dy-
nasties or reigns of Agassiz, and ages of Dana ; the reign
of mollusks, the reign of fishes, of reptiles, of mammals,
and finally of *man.* During the early Palæozoic times
(Cambrian and Silurian) there were no vertebrates.†
But never in the history of the earth were mollusks of
greater size, number, and variety of form than then.

---

* Genesis without previous life—spontaneous generation.

† Fishes were first introduced in the later Silurian ; but became
dominant in the Devonian.

They were truly the rulers of these early seas. In the absence of competition of still higher animals, they had things all their own way, and therefore grew into a great monopoly of power. In the later Palæozoic (Devonian) fishes were introduced. They increased rapidly in size, number, and variety; and being of higher organization they quickly usurped the empire of the seas, while the mollusca dwindled in size and importance, and sought safety in a less conspicuous position. In the Mesozoic times, reptiles, introduced a little earlier,* finding congenial conditions and an unoccupied place above, rapidly increased in number, variety, and size, until sea and land seem to have swarmed with them. Never before or since have reptiles existed in such numbers, in such variety of form, or assumed such huge proportions; nor have they ever since been so highly organized as then. They quickly became rulers in every realm of Nature—rulers of the sea, swimming reptiles; rulers of the land, walking reptiles; and rulers of the air, flying reptiles. In the unequal contest, fishes therefore sought safety in subordination. Meanwhile mammals were introduced in the Mesozoic, but small in size, low in type (marsupials), and by no means able to contest the empire with the great reptiles. But in the Cenozoic (Tertiary) the conditions apparently becoming favorable for their development, they rapidly increased in number, size, variety, and grade of organization, and quickly overpowered the great reptiles,

---

* Amphibians were introduced in the Carboniferous, but true reptile not until the Permian.

which almost immediately sank into the subordinate position in which we now find them, and thus found comparative safety. Finally, in the Quaternary, appeared
man, contending doubtfully for a while, with the great
mammals, but soon (in Psychozoic) acquiring mastery
through superior intelligence. The huge and dangerous
mammals were destroyed and are still being destroyed;
the useful animals and plants were preserved and made
subservient to his wants; and all things on the face of
the earth are being readjusted to the requirements of his
rule. In all cases it will be observed that the rulers were
such because, by reason of strength, organization, and intelligence, they were fittest to rule. There is always room
at the top. To illustrate again by a growing tree: This
successive culmination of higher and higher classes may
be compared to the flowering and fruiting of successively
higher and higher branches. Each uppermost branch,
under the genial heat and light of direct sunshine, received in abundance by reason of position, grew rapidly,
flowered, and fruited; but quickly dwindled when overshadowed by still higher branches, which, in their turn,
monopolized for a time the precious sunshine.

But observe, furthermore : when each ruling class declined in importance, it did not perish, but continued in
a subordinate position. Thus, the whole organic kingdom became not only higher and higher in its highest
forms, but also more and more complex in its structure
and in the interaction of its correlated parts. The whole
process and its result is roughly represented in the ac-

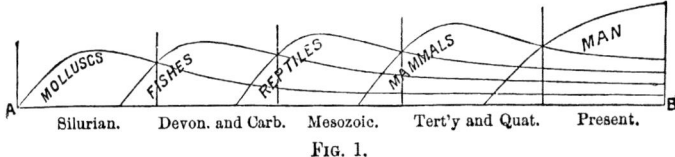

Silurian.   Devon. and Carb.   Mesozoic.   Tert'y and Quat.   Present.

FIG. 1.

companying diagram, Fig. 1, in which A B represents the course of geological time and the curve, the rise, culmination, and decline of successive dominant classes.

THE ABOVE THREE LAWS ARE LAWS OF EVOLUTION.

These three laws we have shown are distinctly recognizable in the succession of organic forms in the geological history of the earth. They are, therefore, undoubtedly the *general laws of succession*. Are they also laws of evolution ? Are they also discoverable in embryonic development, the type of evolution ? They are, as we now proceed to show :

Differentiation.—In reproduction the new individual appears : 1. As a *germ*-cell—a single microscopic living cell. 2. Then, by growth and multiplication of cells, it becomes an *egg*. This may be characterized as an aggregate of *similar* cells, and therefore is not yet differentiated into tissues and organs. In other words, it is not yet visibly organized ; for organization may be defined as the possession of different parts, performing different functions, and all co-operating for one given end, viz., the life and well-being of the organism. 3. Then commences the really characteristic process of development, viz., *differentiation* or diversification. The cells are at first all alike in form and function, for all are globular in form, and each performs all the functions necessary for life. From this common point now commences development in *different directions*, which may be compared to a branching and rebranching, with more and more complex results, according as the animal is higher in the scale of organization and advances toward a state of maturity. First, the cell-aggregate (egg) separates into three distinct layers of cells, called ecto-blast, endo-blast, and meso-blast. These by further differentiation form the three fundamental groups of organs and functions, viz., the *nervous system*, the *nutritive system*, and the *blood system :* the first presiding over the exchange of *force* or influence, by action and reaction with the environment, and between the different parts of the organism ; the second presiding over the exchange of *matter* with the environment, by absorption and elimination ; the third presiding over exchanges of matter between different parts of the organism. The first system of functions and organs may be compared to a system of telegraphy, foreign and domestic ; the second to foreign commerce ; the third to an internal carrying-trade. Following out any one of these groups in higher animals, say the nervous system, it quickly differentiates again into two sub-systems, viz., cerebro-spinal and ganglionic, each having its own distinctive functions, which we can not stop to explain. Then the cerebro-spinal again differentiates into voluntary and reflex systems. All of these have meanwhile separated into sensory and motor centers and fibers. Then, taking only the sensory fibers, these again are differentiated into five special senses, each having a wholly different function. Then, finally, taking any one of these, say the *sense of touch* or feeling, this again is differentiated into many kinds of fibers, each responding to a different impression, some to heat, others to cold, still others to pressure, etc. We have taken the nervous system ; but the same differentiation and redifferentiation takes place in all other systems, and is carried to higher and higher points according to the position in the scale of the animal which is to be formed.

Or, to vary the mode of presentation a little, the cells of the original aggregate, commencing all alike, immediately begin to take on different forms, in order to perform different functions. Some cells take on a certain form and aggregate themselves to form a peculiar tissue which we call muscle, and which does nothing else, can do nothing else, than contract under stimulus. Another group of cells take on another peculiar form and aggregate themselves to form another and very different tissue, viz., nervous tissue, which does nothing and can do nothing but carry influence back and forth between the great external world and the little world of consciousness within. Still another group of cells take still another form and aggregate to form still another tissue, viz., the *epithelial,* whose only function is to absorb nutritive and eliminate waste matters. Thus, by differentiation of form and limitation of function, or division of labor, the different parts of the organism are bound more and more closely together by mutual dependence, and the whole becomes more and more distinctly individuated, and separation of parts becomes more and more a mutilation, and finally becomes impossible without death. This process, as already said, reaches its highest point only in the later stages of development of the highest animals.

Progress.—The *law of progress* is, of course, admitted to be a law of ontogeny ; but observe here, also, it is true only of the whole and not necessarily of all the parts, *except from the point of view of the whole.* Thus, for example, starting all from a common form or generalized type, some cells *advance* to the dignity of brain-cells, whose function is somehow connected with the generation or at least the manifestation of thought, will, and emotion ; other cells *descend* to the position of kidney-cells, whose sole function is the excretion of urine. But here, also, the highest cells are successively higher, and the whole aggregate is successively nobler and more complex. It is again a branching and rebranching, in every direction, some going upward, some downward, some horizontally, anywhere, everywhere, to increase the complexity of relations internal and external, and therefore to elevate the plane of the whole.

Cyclical Movement.—Lastly, the law of cyclical movement is also a law of ontogeny and therefore of evolution. This law, however, is less fundamental than the other two, and is, therefore, less conspicuous in the ontogenic than in the phylogenic series. It is conspicuous only in the later stages of ontogeny, and in other higher kinds of evolution, such as social evolution. For example, in the ontogenic development of the body and mind from childhood to manhood we have plainly successive culminations and declines of higher and higher functions. In bodily development we have culminating first the *nutritive* functions, then the *reproductive* and *muscular*, and last the *cerebral*. In mental development we have culmination first of the receptive and retentive faculties in childhood, then of imaginative and æsthetic faculties in youth and young manhood ; then of the reflective and elaborative faculties—the faculties of productive work in mature manhood ; and, finally, the moral and religious sentiments in old age. The first gathers and stores materials ; the second vivifies and makes them plastic building materials ; the third uses them in actual constructive work—in building the temple of science and philosophy ; and the fourth dedicates that temple only to noblest purposes.

Observe here, also, that when each group of faculties culminates and declines, it does not perish, but only becomes subordinate to the next higher dominant group, and the whole psychical organism becomes not only higher and higher in its highest parts, but also more and more complex in its structure and in the interaction of its correlated parts.

Observe, again, the necessity laid upon us by this law— the necessity of continued evolution to the end. Childhood, beautiful childhood, can not remain — it must

quickly pass. If, with the decline of its characteristic
faculties, the next higher group characteristic of youth
do not increase and become dominant, then the glory of
life is already past and deterioration begins.   Have we
not all seen sad examples of this ?   Youth, glorious
youth, must also pass.   If the next higher group of reflective and elaborative faculties do not arise and dominate, then progressive deterioration of character commences here—thenceforward the whole nature becomes
coarse, as we so often see in young men, or else shrivels
and withers, as we so often see in young women.   Finally, manhood, strong and self-relying manhood, must also
pass.   If the moral and religious sentiments have not
been slowly growing and gathering strength all along,
and do not now assert their dominance over the whole
man, then commences the final and saddest decline of
all, and old age becomes the pitiable thing we so often
see it.   But, if the evolution have been normal throughout ; if the highest moral and religious nature have been
gathering strength through all, and now dominates all,
then the psychic evolution rises to the end—then the
course of life is like a wave rising and cresting only at
the moment of its dissolution, or, like the course of the
sun, if not brightest at least most glorious in its setting.
And thus—may we not hope ?—the glories of the close of
a well-spent life become the pledge and harbinger of an
eternal to-morrow ?

We have thus far illustrated the three laws of succession of organic forms by ontogeny, because this is the

type of evolution ; but they may be illustrated also by other forms of evolution. Next to the development of the individual, undoubtedly the *progress of society* furnishes the best illustration of these laws.

Commencing with a condition in which each individual performs all necessary social functions, but very imperfectly ; in which each individual is his own shoemaker and tailor, and house-builder and farmer, and therefore all persons are socially alike ; as society advances, the constituent members begin to diverge, some taking on one social function and some another, until in the highest stages of social organization this diversification or division and subdivision of labor reaches its highest point, and each member of the aggregate can do perfectly but one thing. Thus, the social organism becomes more and more strongly bound together by mutual dependence, and separation becomes mutilation. I do not mean to say that this extreme is desirable, but only that an approach to this is *a* natural law of social development. *Is not this the law of differentiation?*

So also *progress* is here, as in other forms of evolution—a *progress of the whole*, but not necessarily of every part. Some members of the social aggregate advance *upward* to the dignity of statesmen, philosophers, and poets ; some advance *downward* to the position of scavengers and sewer-cleansers.* But the highest members are progressively higher, and the whole aggregate is

---

* Of course I mean downward in *social function*. Individually the scavenger may be nobler than the statesman.

progressively grander and more complex in structure and functions.

So, again, the *law of cyclical movement* is equally conspicuous here. Society everywhere advances, not uniformly, but by successive waves, each higher than the last ; each urged by a new and higher social force, and embodying a new and higher phase of civilization. Again : as each phase declines, its characteristic social force is not lost, but becomes incorporated into the next higher phase as a subordinate principle, and thus the social organism as a whole becomes not only higher and higher, but also more and more complex in the mutual relations of its interacting social forces.

Let us not be misunderstood, however. There is undoubtedly in social evolution something more and higher than we have described, but which does not concern us here, except to guard against misconstruction. There is in society a *voluntary progress* wholly different from the evolution we have been describing. In *true* or material evolution natural law works for the betterment of the whole utterly regardless of the elevation of the individual, and the individual contributes to the advance of the whole quite unconsciously while striving only for his own betterment. This unconscious evolution by natural law inherited from the animal kingdom is conspicuous enough in society, especially in its early stages, but we would make a great mistake if we imagined, as some do, that this is all. Besides the unconscious evolution by natural laws, *inherited from below*, there is a higher evolution, *inherited from above,* indissolubly connected with man's spiritual nature—a conscious, voluntary striving of the best members of the social aggregate for the betterment of the whole—a conscious, voluntary striving both of the individual and of society toward a recognized *ideal.* In the one kind of evolution the fittest are those most in harmony with the environment, and which therefore always survive ; in the other, the fittest are those most in harmony with the ideal, and which often do not survive. The laws of this free voluntary progress are little understood. They are of supreme importance, but do not specially concern us here. We will speak of it again in another chapter.

The three laws above mentioned might be illustrated equally well by all other forms of evolution. We have selected only those which are most familiar. They may, therefore, be truly called the laws of evolution. We have shown that they are the laws of succession of organic forms.

III. **Change by Means of Resident Forces.**—Thus far in our argument I suppose that most well-informed men will raise no objection. It will be admitted, I think, even by those most bitterly opposed to the theory of evolution, that there has been throughout the whole geological history of the earth an onward movement of the organic kingdom to higher and higher levels. It will be admitted, also, that there is a grand and most significant resemblance between the course of development of the organic kingdom and the course of embryonic develop-

4

ment—between the laws of succession of organic forms
and the laws of ontogenic evolution. But there is another essential element in ontogenic evolution. It is
that the *forces* or causes of evolution are *natural ;* that
they reside in the thing developing and in the reacting
environment. This we know is true of embryonic development ; is it true also of the geologic succession of organic forms ? It is true of ontogeny ; is it true also of
phylogeny ? If not, then only by a metaphor can we call
the process of change in the organic kingdom throughout
geological history an evolution. This is the point of
discussion, and not only of discussion, but, alas ! of
heated and even angry dispute. The field of discussion
is thus narrowed to this third point only.

Before stating the two opposite views of the cause of
evolution, it is necessary to remind the reader that when
the evolutionist speaks of the forces that determine progressive changes in organic forms as *resident* or *inherent*,
all that he means, or ought to mean, is that they are
resident in the same sense as all natural forces are resident ; in the same sense that the vital forces of the embryo are resident in the embryo, or that the forces of the
development of the solar system according to the nebular
or any other cosmogonic hypotheses are resident in that
system. In other words, they mean only that they are
natural, not supernatural. This does not, of course,
touch that deeper, that deepest of all questions, viz.,
the essential *nature and origin of natural forces ;* how
far they are independent and self-existent, and how far

they are only modes of divine energy. This is a question of philosophy, not of science. This question is briefly discussed in another place (Part III, Chap. III) ; it does not immediately concern us here.

The Two Views briefly Contrasted.—As already stated, all will admit a grand resemblance between the stages of embryonic development and those of the development of the organic kingdom. This was first brought out clearly by Louis Agassiz, and is, in fact, the greatest result of his life-work. All admit, also, that the embryonic development is a natural process. Is the development of the organic kingdom also a natural process ? All biologists of the present day contend that it is ; all the old-school naturalists, with Agassiz at their head, and all anti-evolutionists of every school, contend that it is not. We take Agassiz as the type of this school, because he has most fully elaborated and most distinctly formulated this view. As formulated by him, it has stood in the minds of many as an alternative and substitute for evolution.

According to the evolutionists, all organic forms, whether species, genera, families, orders, classes, etc., are variable, and, if external conditions favor, these variations accumulate in one direction and gradually produce new forms, the intermediate links being usually destroyed or dying out. According to Agassiz, the higher groups, such as genera, families, orders, etc., are indeed variable by the introduction of new species, but species are the ultimate elements of classification, and, like the ultimate elements of chemistry, are unchangeable ; and, therefore, the speculations of the evolutionist concerning the transmutation of species are as vain as were the speculations of the alchemists concerning the transmutation of metals—that the origin of man, for example, from any lower species is as impossible as the origin of gold from any baser metal. Both sides admit frequent change of species during geological history, but one regards the change as a change by gradual *transmutation* of one species *into* another through successive generations and by *natural* process, the other as change by *substitution* of one species *for* another by direct supernatural *creative act*. Both admit the gradual development of the organic kingdom as a whole through stages similar to those of embryonic development ; but the one regards the whole process as natural, and therefore strictly comparable to embryonic development, the other as requiring frequent special interference of creative energy, and therefore comparable rather to the development of a building under the hand and according to the preconceived plan of an architect—a plan, in this case, conceived in eternity and carried out consistently through infinite time. It is seen that the essential point of difference is this : The one asserts the variability of species (if conditions favor, and time enough is given) without limit ; the other asserts the permanency of specific forms, or their variability only within narrow limits. The one asserts the origin of species by "*descent with modifications*" ; the other, the origin of species by "*special act of creation.*" The one asserts the law of continuity (i. e., that each stage is the natural outcome of the immediately preceding stage) in this, as in every other department of Nature; the other asserts that the law of continuity (i. e., of cause and effect) does not hold in this department; that the links of the chain of changes are discontinuous, the connection between them being intellectual, not physical.

So much for sharp contrasting characterization of the two views, necessary for clear understanding of much that follows. We will have to give them more fully hereafter when we take up the evidences of evolution in Part II.

CHAPTER II.

IN order to clear up the conception of evolution, it is
necessary to give a brief history of the idea, and especially to explain the relation of Louis Agassiz to that
theory. This is the more necessary, because there is a
deep and wide-spread misunderstanding on this subject,
and thus scant justice has been done our great naturalist,
especially by the English and Germans ; and also because
this relation is an admirable illustration of an important
principle in scientific philosophy.

Like all great ideas, we find the first germs of this in
Greek philosophy, in the cosmic speculations of Thales
and Pythagoras. Next (about 100 B. C.) we find it more
clearly expressed by the Roman thinker, Lucretius, in
his great philosophic poem entitled "De Rerum Natura."
After a dormancy of nearly eighteen centuries it next
emerges with still more clearness in the theological speculations of Swedenborg and the philosophical speculations
of Kant. All these we pass over with bare mention, because these thinkers approached the subject from the

philosophic rather than the scientific side—in the meta-physical rather than the scientific spirit.

The first serious attempt at scientific presentation of the subject was by the celebrated naturalist, Lamarck, in a work entitled "Philosophie Zoölogique," published in 1809. It is not necessary, in this rapid sketch, to give a full account of Lamarck's views. Suffice it to say that the essential idea of evolution, viz., the indefinite variability and the derivative origin of species, was insisted on with great learning and skill, and illustrated by many examples. With Lamarck, the factors of evolution or causes of change of organic forms were—1. Modification of organs in function and therefore in structure, by a changing environment—external factor ; and, 2. Modification of organs by *use* and *disuse*—internal factor. In both cases the modifications are inherited and increased from generation to generation, without limit. This second factor seems to have taken, in the mind of Lamarck, the somewhat vague and transcendental form of aspiration or upward striving of the animal toward higher conditions. These are acknowledged to-day as true factors of evolution, but the distinctively Darwinian factor, viz., "divergent variation and natural selection," was not then thought of. The publication of Lamarck's views produced a powerful impression, but only for a little while. Pierced by the shafts of ridicule shot by nimble wits of Paris, and crushed beneath the heavy weight of the authority of Cuvier, the greatest naturalist and comparative anatomist of that or perhaps

of any time, it fell almost still-born. I believe it was
best that it should thus perish. Its birth was prema-
ture ; it was not fit to live. The world was not yet prepared for a true scientific theory. Nevertheless, the
work was not without its effect upon some of the most
advanced thinkers of that time ; upon Saint-Hilaire and
Comte in France, and upon Goethe and Oken in Ger-
many. It was good seed sown and destined to spring up
and bear fruit in suitable environment ; but not yet.

The next attempt worthy of attention in this rapid
sketch is that of Robert Chambers, in a little volume entitled "Vestiges of a Natural History of Creation," published in 1844. It was essentially a reproduction of
Lamarck's views in a more popular form. It was not a
truly scientific work nor written by a scientific man. It
was rather an appeal from the too technical court of science to the supposed wider and more unprejudiced court
of popular intelligence. It was therefore far more elo-
quent than accurate ; far more specious than profound.
It was, indeed, full of false facts and inconsequent reasonings. Nevertheless, it produced a very strong impression on the thinking, popular mind. But *it* also quickly
fell, pierced by keen shafts of ridicule, and crushed beneath the heavy weight of the authority of all the most
prominent naturalists of that time, with Agassiz at their
head. The question for the time seemed closed. I believe, again, it was best so, for the time was not yet fully
ripe.

I know full well that many think with Haeckel that

biology was kept back half a century by the baneful authority of Cuvier and Agassiz ; but I can not think so.
The hypothesis was contrary to the facts of science *as
then known and understood*. It was conceived in the
spirit of baseless speculation, rather than of cautious
induction ; of skillful elaboration rather than of earnest
truth-seeking. Its general acceptance would have de-
bauched the true spirit of science. I repeat it : the time
was not yet ripe for a scientific theory. The ground
must first be cleared and a solid foundation built ; an insuperable *obstacle* to hearty rational acceptance must
first be removed, and an inductive *basis* must be laid.

The Obstacle removed.—The obstacle in the way of
the acceptance of the derivative origin of species was the
then prevalent *notion concerning the nature of life*. We
must briefly sketch the change which has taken place in
the last forty years in our ideas on this subject.

Until about forty years ago, the different forces of
Nature, such as gravity, electricity, magnetism, light,
heat, chemical affinity, etc., were supposed to be entirely
distinct. The realm of Nature was divided up into a
number of distinct and independent principalities, each
subject to its own sovereign force and ruled by its own
petty laws. About that time it began to be evident, and
is now universally acknowledged, that all these forces are
but different *forms* of one, universal, omnipresent energy,
and are transmutable unto one another back and forth
without loss. This is the doctrine of correlation of
forces and conservation of energy, one of the grandest

ideas of modern times. But *one* force seemed still to be
an exception. Life-force was still believed to be a peculiar, mysterious principle or entity, standing above
other forces and subordinating them ; not correlated
with, not transmutable unto, nor derivable from, other
and lower forces, and therefore in some sense supernatural. Now, if this be true of living *forces*, it is perfectly natural, yea, almost necessary, to believe that living *forms* are wholly different from other forms in their
origin. New forms of dead matter may be derived, but
new living forms are *underived*. Other new forms come
by natural process, new organic forms by supernatural
process. The conclusion was almost unavoidable. But
soon vital force also yielded to the general law of correlation of natural forces. Vital forces are also transmutable
into and derivable from physical and chemical forces.
Sun-force, falling on the green leaves of plants, is absorbed and converted into vital force, disappears as *light*
to reappear as *life*. The amount of life-force generated
is measured by the amount of light extinguished. The
same is true of animal life. As in the steam-engine the
locomotive energy is derived from the fuel consumed and
measured by its amount, so in the animal body, the animal heat and animal force are derived from and measured
by the food and tissue consumed·by combustion. Thus,
vital force may be regarded as so much force withdrawn
from the general fund of chemical and physical forces, to
be again refunded without loss at death. This obstacle
is, therefore, now removed. If vital force falls in the

same category as other natural forces, there is no reason why living forms should not fall into the same category in this regard as other natural forms. If new forms of dead matter are derived from old forms by modification, according to *physical* laws, there is no reason why new living forms should not also be derived from old forms by modification according to *physiological* laws. Thus, at last, the obstacle was removed—the ground was cleared.

The Basis laid.—But Science is not content with removal of *a priori* objections. She must also have positive proofs. The ground must not only be cleared, but a true inductive basis of facts, and especially of laws and methods, must be laid. *This was the life-work of Agassiz.* Yes, as strange as it may seem to some, it is nevertheless true that the whole inductive basis, upon which was afterward built the modern theory of evolution, was laid by Agassiz, although he himself persistently refused to build upon it any really scientific superstructure. It is plain, then, that all attempts at building previous to Agassiz's work must, of necessity, have resulted in an unsubstantial structure—an edifice built on sand, which could not and ought not to stand. I must stop here in order to explain somewhat fully this important point, and thus to give due credit to the work of Agassiz.

The title of any scientist to greatness must be determined, not so much by the multitude of new facts he has discovered as by the new laws he has established, and especially by the new methods he has inaugurated or per-

fected. Now, I think it can be shown that to Agassiz,
more than to any other man, is due the credit of having
established the laws of succession of living forms in the
geological history of the earth—laws upon which must
rest any true theory of evolution. Also, that to him,
more than to any other man, is due the credit of having
perfected the method (method of comparison) by the use
of which alone biological science has advanced so rapidly
in modern times. This is high praise. I wish to justify
it. I begin with the method.

Scientific methods bear the same relation to *intellectual progress* that tools, instruments, machines, mechanical contrivances of all sorts, bear to *material progress*.
They are intellectual *contrivances*—indirect ways of accomplishing results far too hard for bare-handed, unaided
intellectual strength. As the civilized man has little or
no advantage over the savage in bare-handed strength of
muscle, and the enormous superiority of the latter in
accomplishing material results is due wholly to the use
of mechanical contrivances or machines; even so, in the
higher sphere of intellect, the scientist makes no preten-
sion to the possession of greater unaided intellectual
strength than belongs to the uncultured man, or even
perhaps to the savage. The amazing intellectual results
achieved by science are due wholly to the use of intellectual contrivances or scientific methods. As in the lower
sphere of material progress the greatest benefactors of
the race are the inventors or perfecters of new mechanical contrivances or *machines*, so also in the higher sphere

of intellectual progress the greatest benefactors of the race are the inventors or perfecters of new intellectual contrivances or *methods of research.*

To illustrate the power of methods, and the necessity of their use, take the case of the *method of notation,* so characteristic of mathematics, and take it even in its simplest and most familiar form : Nine numeral figures, having each a value of its own, and another dependent upon its position ; a few letters, *a* and *b,* and *x* and *y,* connected by symbols, + and — and = : that is all.   And yet, by the use of this simple contrivance, the dullest school-boy accomplishes intellectual results which would defy the utmost efforts of the unaided strength of the greatest genius.   And this is only the simplest tool-form of this method.   Think of the results accomplished by the use of the more complex machinery of the higher mathematics !

Take next the method of experiment so characteristic of physics and chemistry.   The phenomena of the external world are far too complex and far too much affected by disturbing forces and modifying conditions to be understood at once by bare, unaided intellectual insight.   They must first be simplified.   The physicist, therefore, contrives artificial phenomena under ideal conditions.   He removes one complicating condition after another, one disturbing cause and then another, watching meanwhile the result, until finally the necessary condition and the true cause are discovered.   On this method rests the whole fabric of the physical and chemical sciences.

But when we rise still higher, viz., into the plane of life, the phenomena of Nature become still more complex and difficult to understand directly; and yet just here, where we are the most powerless without some method, our method of experiment almost wholly *fails us.* The phenomena of life are not only far more complex than those of dead matter, but the conditions of life are so nicely adjusted, the equilibrium of forces so delicately balanced, that, when we attempt to introduce our clumsy hands in the way of experiment, we are in danger of overthrowing the equilibrium, of destroying the conditions of the experiment, viz., life; and then the whole problem falls immediately into the domain of chemistry. What shall we do ? In this dilemma we find that Nature herself has already prepared for us, ready to hand, an elaborate series of simplified conditions equivalent to experiments. The phenomena of life are, indeed, far too complex to be at once understood—the problem of life too hard to be solved—in the higher animals; but, as we go down the animal scale, complicating conditions are removed one by one, the phenomena of life become simpler and simpler, until in the lowest microscopic cell or spherule of living protoplasm we finally reach the simplest possible expression of life. The equation of life is reduced to its simplest terms, and now, if ever, we begin to understand the true value of the unknown quantity. This is the natural history series, or *Taxonomic* series, already spoken of on page 10. Again, Nature has prepared, and is now preparing daily before our eyes, an·

other series of gradually simplified conditions. Commencing with the mature condition of one of the higher animals—for example, man—and going backward along the line of individual history through the stages of infant embryo, egg and germ, we find again the phenomena of life becoming simpler and simpler, until we again reach the simplest conceivable condition in the single microscopic cell or spherule of living protoplasm. This, as already explained, is the embryonic or *Ontogenic* series. Again, that there be no excuse for man's ignorance of the laws of life, Nature has prepared still another series ; and this the grandest of all, for it is the cause of both the others. Commencing with the plants and animals of the present epoch, and going back along the track of geological times, through Cenozoic, Mesozoic, Palæozoic, Eozoic, to the very dawn of life—the first syllable of recorded time—and we find again a series of organic forms growing simpler and simpler, until, if we could find the very first, we would undoubtedly again reach the simplest condition in the lowest conceivable forms of life. This, as we have already seen, is the geologic or evolution, or *Phylogenic* series. We have already explained these three series, only in this connection it suits our purpose to take the terms backward.

Now, it is by *comparison* of the terms of each of these series going up and down, and watching the first appearance, the growth, and the perfecting of tissues, organs, functions, and by the comparison of the three series with one another term by term — I say it is wholly by comparison of this kind that biology has in recent times become a true inductive science. This is the *"method of comparison."* It is the great method of research in all those departments which can not be readily managed by the method of experiment. It has already regenerated biology, and is now applied with like success in sociology under the name of *historic method*. Yes; anatomy became scientific only through comparative anatomy, physiology through comparative physiology, and embryology through comparative embryology. May we not add, sociology will become truly scientific only through comparative sociology, and psychology through comparative psychology?

Now, while it is true that this method, like all other methods, has been used, from the earliest dawn of thought, in a loose and imperfect way, yet it is only in very recent times that it has been organized, systematized, perfected, as a true scientific method, as a great instrument of research; and the prodigious recent advance of biology is due wholly to this cause. Now, among the great leaders of this modern movement, Agassiz undoubtedly stands in the very first rank. I must try to make this point plain, for it is by no means generally understood.

Cuvier is acknowledged to be the great founder of comparative anatomy. He it was that first perfected the method of comparison, but comparison only in one series —the *Taxonomic*. Von Baer and Agassiz added to this, comparison in the ontogenic series also, and comparison of these two series with each other, and therefore the application of embryology to the classification of animals. If Von Baer was the first announcer, Agassiz was the first great practical worker by this method. Last and most important of all, in its relation to evolution, Agassiz added *comparison in the geologic or phylogenic series.* The one grand idea underlying Agassiz's whole life-work was the essential identity of the three series, and therefore the light which they must shed on one another. The two guiding and animating principles of his scientific work were—1. That the embryonic development of one of the higher representatives of any group repeated in a general way the terms of the Taxonomic series in the same group, and therefore that embryology furnished the key to a true classification ; and, 2. That the succession of forms and structure in geological times in any group is similar to the succession of forms and structure in the development of the individual in the same group, and thus that embryology furnishes also the key to geological succession. In other words, during his whole life, Agassiz insisted that the laws of embryonic development (ontogeny) are also the laws of geological succession (phylogeny). Surely this is the foundation, the only solid foundation, of a true theory of evolution. It is true that Agassiz, holding as he did the doctrine of permanency of specific types, and therefore rejecting the doctrine of the derivative origin of species, did not admit the causal or natural relation of phylogenic succes-

5

sion to embryonic succession and taxonomic order as we
now believe it—it is true that for him the relation between the three series was an intellectual not a physical
one—consisted in the preordained plans of the Creator,
and not in any genetic connection or inherited property ;
but evidently the first and greatest step was the discovery
of the relation itself, however accounted for. The rest
was sure to follow.

But more. Not only did Agassiz establish the essential identity of the geologic and embryonic succession,
the general similarity of the two series, phylogenic and
ontogenic, but he also announced and enforced all the
formal laws of geologic succession (i. e., of evolution), as
we now know them. These, as already stated and illustrated, are the law of differentiation, the law of progress
of the whole, and the law of cyclical movement, although he did not formulate them in these words. No
true inductive evidence of evolution was possible without
the knowledge of these laws, and for this knowledge we
are mainly indebted to Agassiz. He well knew also that
they were the laws of embryonic development and therefore of evolution ; but he avoided the word evolution, as
implying the derivative origin of species, and used instead the word *development,* though it is hard to see in
what the words differ. Thus, it is evident that Agassiz
laid the whole foundation of evolution, solid and broad,
but refused to build any scientific structure on it ; he refused to recognize the legitimate, the scientifically necessary outcome of his own work. Nevertheless, without his

work a scientific theory of evolution would have been impossible. Without Agassiz (or his equivalent), there would have been no Darwin.

There is something to us supremely grand in this refusal of Agassiz to accept the theory of evolution. The opportunity to become the leader of modern thought, the foremost man of the century, was in his hands, and he refused, because his religious, or, perhaps better, his philosophic intuitions, forbade. To Agassiz, and, indeed, to all men of that time, to many, alas! even now, evolution is materialism. But materialism is Atheism. Will some one say, the genuine Truth-seeker follows where she seems to lead *whatever be the consequences?* Yes; whatever be the consequences to one's self, to one's opinions, prejudices, theories, philosophies, but not to *still more certain truth.* Now, to Agassiz, as to all genuine thinkers, the existence of God, like our own existence, is more certain than any scientific theory, than anything can possibly be made by proof. From his standpoint, therefore, he was right in rejecting evolution as conflicting with still more certain truth. The mistake which he made was in imagining that there was any such conflict at all. But this was the universal mistake of the age. A lesser man would have seen less clearly the higher truth and accepted the lower. A greater man would have risen above the age, and seen that there was no conflict, and so accepted both. All thinking men are coming to this conclusion now, but none had done so then.

Now, then, at last, the obstacle of supernaturalism in the realm of Nature having been removed by the establishment of the doctrine of correlation of natural forces, and the extension of this doctrine to embrace also life-force ; and now also a broad and firm basis of carefully-observed facts and well-established laws of succession of organic forms having been laid by Agassiz, when again, for the third time, the doctrine of origin of species "by derivation with modifications" was brought forward by Darwin in a far more perfect form, with more abundant illustrative materials, and with a new and most potent factor of modification—viz., divergent variations and natural selection—it found the scientific world already fully prepared, and anxiously waiting. I say *anxiously* waiting—for the supposed supernatural origin of species had been the one exception to the otherwise universal law of cause and effect, or the law of continuity. It was therefore in open contradiction to the whole drift of scientific thought for five hundred years. Is it any wonder, then, that the derivative origin of species was welcomed with joy by the scientific world ? For five hundred years, scientific thought, like a rising tide which knows no ebb, had tended thitherward with ever-increasing pressure, but kept back by the one supposed fact of the supernatural origin of species. Darwin lifted the gate, and the in-rushing tide flooded the whole domain of thought.

What, then, is the place of Agassiz in biological science ? What is the relation of Agassiz to Darwin—of

Agassizian development to Darwinian evolution ?   I answer, it is the relation of formal science to physical or causal science.   Agassiz advanced biology to the *formal* stage ; Darwin carried it forward, to some extent at least, to the *physical* stage.   All true inductive sciences in their complete development pass through these two stages.   Science in the one stage treats of the *laws* of phenomena ; in the other, of the *causes* or explanation of these laws.   The former must precede the latter, and form its foundation ; the latter must follow the former, and constitute its completion.   The change from the one to the other is always attended with prodigious impulse to science.

To illustrate : Until Kepler, astronomy was little more than an accumulation of disconnected facts concerning celestial motions—abundant materials, but no science ; piles of brick and stone, but no building.   Kepler reduced this chaos to beautiful order and musical harmony by the discovery of the three great laws which bear his name, and therefore he has been justly called the legislator of the heavens—*the lawgiver of space*.   But, had he been asked the *cause* of these beautiful laws, he could only have answered, "The *first cause*—the direct will of the Deity."   A good answer and a true, but not scientific ; because it places the question beyond the domain of science, which deals only with second or physical causes.   But Newton comes forward and gives a *physical cause*.   He shows that all these beautiful laws are the necessary result of gravitation ; and thus astronomy be-

comes a physical science.   So, until Agassiz, the facts of
geological succession of organic forms were in a state of
lawless confusion.   Agassiz by establishing the three great
laws of succession, which ought to bear his name, reduced this chaos to order and beauty ; and, therefore, he
might justly be called the legislator of geological history
—the *lawgiver of time*.   But, when asked the cause of
these laws, he could only answer, and did indeed answer, " The plans of the Creator."   A noble answer and
true, but not scientific.   Darwin now comes forward and
gives, partly at least, the cause of these laws.   He shows
that all these beautiful laws are explained by the doctrine of " origin of species by derivation with modifications " ; that these laws are not ultimate, but derivative
from more fundamental laws of life ; and thus biology is
advanced one step, at least, toward the causal stage.
Newton and Darwin substituted second causes for first
cause—natural for supernatural.   They each in his own
department broke the bonds of supernaturalism in the
domain of Nature.

One more important reflection : There are two, and
only two, fundamental conditions of material existence—
space and *time*.   There are, therefore, two, and only two,
cosmoses — space-cosmos and time-cosmos.   These have
been redeemed from confusion and reduced to law and
order and beauty—changed from chaos to cosmos—by
science.   For this result we are chiefly indebted, in the
one case, to Kepler and Newton ; in the other, to Agassiz
and Darwin.   The universal law, in the one cosmos, is

the *law of gravitation;* in the other, the *law of evolution.*
Traced by analysis to its deepest roots of philosophic
truth, the one law may be called the divine mode of sustentation ; the other, the divine process of creation.

Or again : we have all heard of the "music of the
spheres"—a beautiful and significant name used by the
old thinkers for the divine order of the universe—a
music heard not by human ear, but only by the atten-
tive human spirit.   Harmonic relation apprehended by
reason we call *Law,* and its embodiment Science ; the
same apprehended by the imagination and æsthetic
sense, we call *Beauty,* and its embodiment *Art, music.*
Now, in music there are two kinds of harmony, simul-
taneous and consecutive—chordal harmony and melody.
These must be combined to produce the grandest effect.
So in cosmic order, too, there are two kinds of harmonic
relation—the *co-existent in space* and the *consecutive in
time.*   The law of gravitation expresses the universal
harmonic inter-relation of *objects* co-existent in space,
the law of evolution, the universal harmonic relation of
forms successive in time.   Of the divine spheral music,
the one is the chordal harmony, the other the consecutive
harmony or melody.   Combined they form the divine
chorus which "the morning stars sang together."

PART II.

EVIDENCES OF THE TRUTH OF EVOLUTION.

CHAPTER I.

LET us again remind the reader that evolution means, first of all, *continuity*. The law of evolution, although it doubtless means much more, means, first of all, a law of continuity, or *causal relation throughout Nature*. It means that, alike in every department of Nature, each state or condition grew *naturally* out of the immediately preceding. In a word, it means that, in the course of Nature, nothing appears suddenly and without natural cause, but, on the contrary, everything is the natural and usually the gradual outcome of a previous condition. This is *now* admitted by every one in regard to *nearly* everything : evolutionists apply it to the whole course of Nature. I said this is *now* admitted by every one in regard to *nearly* everything ; but this has not always been so. The world has come to its present position on this subject only by a very gradual process. Let us then trace rapidly the history of the gradual change, for it will prepare us for much that follows.

There was a time (and that not many decades ago)

when all things, the origin of which transcends our ordinary experience, were supposed to have originated suddenly and without natural process—to have been made at once, out of hand. There was a time when, for example, mountains were supposed to have been made at once, with all their diversified forms, of beetling cliffs and thundering waterfalls, or gentle slopes and smiling valleys, just as we now find them. But *now* we know that they have become so only by a very gradual process, and are still changing under our very eyes. In a word, they have been formed by a *process of evolution.* We know now the date of mountain-births; we trace their growth, maturity, decay, and death; and find even, as it were, the fossil bones of extinct mountains in the crumpled strata of their former places. There was a time when continents and seas, gulfs, bays, and rivers, were supposed to have originated at once, substantially as we now see them. *Now,* we know that they have been changing throughout all geological time, and are still changing. Not, however, change back and forth in any direction indifferently and without goal, but gradual change from less perfect to more perfect condition, with more and more complex inter-relations—i. e., by a *process of evolution.* We are able now, though still imperfectly, to trace some of the stages of this evolution. There was a time when rocks and soils were supposed to have been always rocks and soils; when soils were regarded as an original clothing made on purpose to hide the rocky nakedness of the new-born earth. God clothed the earth so, and there an end. *Now* we know that rocks rot down to soils ; soils are carried down and deposited as sediments ; and sediments reconsolidate as rocks—the same materials being worked over and over again, passing through all these stages many times in the history of the earth. In a word, there was a time when it was thought that the earth with substantially its present form, configuration, and climate, was made at once out of hand, as a fit habitation for man and animals. *Now* we know that it has been changing, preparing, becoming what it is by a slow process, through a lapse of time so vast that the mind sinks exhausted in the attempt to grasp it. It has become what it now is by a *process of evolution.* The same change of view has taken place concerning the origin of all the heavenly bodies. We may, therefore, confidently generalize—we may assert without fear of contradiction that *all inorganic forms,* without exception, have originated by a process of evolution.

The proof of all this we owe to geology—a science born of the present century. This science establishes the law of *universal continuity* of events, through infinite *time,* as astronomy does that of *universal inter-relation* of objects through infinite *space.* How great the change these two sciences have made in the realm of human thought ! Until the birth of modern astronomy the intellectual *space-horizon* of the human mind was bounded substantially by the dimensions of our earth ; sun, moon, and stars, being but inconsiderable bodies circulating at a little distance about the earth, and for our behoof. Astronomy was then but the geometry of the curious lines traced by these wandering fires on the concave blackboard of heaven. With the first glance through a telescope the phases of Venus and the satellites of Jupiter, revealed clearly to the mind the existence of other worlds besides and like our own. In that moment the idea of *infinite space*, full of worlds like our own, was for the first time completely realized, and became thenceforward the heritage of man. In that moment the *intellectual horizon of man was infinitely extended*. So also until the birth of geology, about the beginning of the present century, the intellectual *time-horizon* of the human mind was bounded by six thousand years. The discovery about that time of vertebrate remains, all wholly different from those now inhabiting the earth, revealed the existence of other time-faunas, besides our own and the idea of infinite time, of which the life of humanity is but an epoch, was born in the mind of man ; and again the intellectual horizon of man was infinitely extended. These two are the grandest ideas, and their introduction the grandest epochs, in the intellectual history of man. We have long ago accepted and readjusted our mental furniture to the requirements of the one, but the necessary readjustment to the other is not yet complete.

All inorganic forms, then, it is admitted, have come by evolution. But how is it with organic or living forms ? Let us see.

Every one knows, because it is within the limits

of ordinary experience, that every *individual* organism *now* originates and gradually becomes what we see it, by a natural process—that is, by evolution. If, then, there be any exception, it must be only the *first of each kind*. But what kind ? There are many kinds of kinds ; classes, orders, families, genera, species, varieties. Now, many of these kinds can be shown to have become what we see them by a gradual process similar, at least, to evolution. Take for example, classes. The class of fishes and the class of reptiles are *now* widely distinct and have little in common except a vertebrate structure ; but, as already shown, page 12, this extreme difference has not always existed. On the contrary, the earliest representatives of these two classes so merged into one another that each seemed either. From this common stock the two classes were gradually separated, each going its own way and becoming more and more widely distinct even to the present day. There can be no doubt, therefore, that *these two classes*, as we now know them, *have become* what they are by a gradual process. Again : In the whole realm of Nature there is not a class more distinctly separate from every other and without intermediate links than birds. But this has not always been so. They have gradually become so. The earliest birds were so reptilian in structure and appearance that if we could see them now we would be in doubt whether we should call them birds or reptiles. Birds have gradually separated themselves from the reptilian stem, becoming more and more bird-like from age to age, until now, at last, the two classes are wholly separated and the intermediate links destroyed. So far as external characters are concerned, birds may be said to have finally and wholly released themselves from entangling alliance with any other class.

Classes, then, it will be admitted, have undoubtedly become what we now know them by a very gradual process following laws identical (as we have already seen, page 19) with the laws of evolution. Shall we try orders? Of the class Mammalia there are two well-recognized and widely-distinct *orders*, viz., the Carnivores and the Herbivores. We all know how widely diverse these are in form, in structure, in habits, and in food. Has it always been so? Have these been made so at once? By no means. They have gradually become so. The earliest mammals were neither the one nor the other distinctively. They were *omnivores*, completely intermediate in food, habits, form, and structure. From this common stock the two orders have gradually separated, the carnivores becoming more and more adapted to one mode of life and the herbivores to another, by a process following the laws of evolution, as already explained. Shall we try *families* and *genera?* Marsh and Huxley have shown us how completely the horse family (*Equidæ*) and the horse-genus (*Equus*) illustrate the process of gradual becoming and the law of evolution. Under their guidance, we see that the earliest traceable ancestor of the horse family, before it was distinctively a horse family at all, had on the fore-foot five toes in the Lower Eocene, four toes in the Upper Eocene, and three toes in the Miocene ; then we see the two side-toes shortening up more and more in the Pliocene and becoming rudimentary splints, leaving only one toe in the Quaternary and present epochs. Thus, the side-splints in the foot of the modern horse tell the story of its three-toed ancestry. Similar gradual changes are clearly traceable in size, shape, structure of limbs, of teeth, and of brain. In all respects the members of the horse family have become more and more horse-like in the course of time.

This subject will be taken up and more fully illustrated, under the head of special evidences, in a subsequent chapter. We here touch it only sufficiently to illustrate this universal law of gradual becoming.

We have taken only a few examples, but the same is undoubtedly true of all Taxonomic groups *above species.* Passing over these last for the moment, we take next *races* and *varieties.* These smaller groups are admitted by all to be formed by a natural process, because not only can we make them artificially, but all the intermediate links may be found in Nature. So we have only *species* remaining. Yes; species are imagined by the old-school naturalist and by the anti-evolutionist of to-day as the *ultimate elements* of Taxonomy. This, then, is the *last ditch* upon which the defense of supernaturalism in the realm of Nature is made. "Other groups," they say, "may have gradually become what they now are by the successive introduction of specific forms according to a preordained plan which is well expressed by the formal laws of evolution. But *species* are without transition

6

forms. *They* come in suddenly, remain unchanged while they continue, and finally pass out suddenly, so far as specific characters are concerned. New species come in their places by direct act of creation—by *substitution*, not by transmutation." This, then, is the last intrench-ment. Can we give any good evidence of gradual forma-tion of species ? I believe we can.

First, then, it is admitted that we can easily make varieties and races artificially. We will not *now* describe the process ; we are all familiar with the results, viz., the varieties of domestic animals and of useful and orna-mental plants ; the extremely different breeds of horses, cattle, sheep, dogs, pigeons, etc. ; of wheat, cabbages, turnips ; of roses, dahlias, etc., etc. No one will doubt that the extreme varieties of any of these, say greyhound and pug, if wild, would be called distinct species, or even distinct genera. We do not call them so, for two reasons : first, because we see them made ; and, second, because we find all intermediate links between them ; and the usual definition of species is that they can not be made, and they have no intermediate links. Thus, then, the question is narrowed down to *wild species.* They say : "We take our stand on these" (surely a very narrow ground for so broad a philosophy). "We defy you to show gradual formation with intermediate links."

Now, in fact, by diligent search such intermediate links between well-recognized species have been found in some cases, especially in birds, on account of their great power of dispersal. Certain forms have long been known

from widely-separated regions, and universally regarded
as distinct species, as distinct as any. Then, by minute
examinations of intermediate regions, a complete series of
intermediate forms has been picked up. This has occurred not only in one case but in many cases, and not
in birds only but in many other classes—examples increase with our increasing knowledge.* The only answer
to such evidence is that *these are not true species.* Now,
see the fallacy lurking here ! They define species as ultimate elements of taxonomy, as distinct and without
intermediate links, and then require us to find such intermediate links; and, finally, when with infinite pains
some such links are found, they say : "Oh ! I see ; we
were mistaken ; they are only varieties ! !" It is true
that naturalists, when intermediate links are found, usually put all together as one species, but this they do
purely for the sake of clearness of definition and description. It is freely admitted by the evolutionist that species are *now* usually distinct and without intermediate
links, these having been destroyed in the struggle for
life. This will be fully explained in another chapter.
It is also freely admitted that although intermediate
links must have existed at one time, their remains are
rarely found. The reason of this will also be explained
hereafter. Nevertheless, in some cases, as already seen,
we do find them still existing. Now, we add that in
some cases, where they no longer exist, we find them in

* Cope, "Science," vol. ii, p. 274, 1883.

the form of fossil remains. The most remarkable example of this is found in the gradual changes in the forms of Planorbis in the fresh-water deposits of Steinheim, as shown by the admirable researches of Hyatt.* We shall discuss these also more fully in another place. Now, if there be any such links at all, however rare, then every objection to the derivative origin of species is removed.

Perhaps it may be well to make bare mention of another kind of evidence, viz., the actual change of species under the eyes, by the action of change of environment. The different species of the genus *Artemia* (a low form of crustacean) live in brine-pools. By concentrating the brine of such a pool, one species (*A. salina*) has been observed to change in successive generations into another (*A. Muhlhausenii*), and the latter back again to the former by slow freshening.† Again : The siredon and the amblystoma have always, until recently, been regarded as not only distinct species, but distinct genera of amphibians. Siredon was supposed to be a permanent gill-breather, while amblystoma becomes by metamorphosis a pure air-breather. Now, however, it is known that the former may change into the latter. But the most curious part of the life-history of these animals, is that if water be abundant the siredon reproduces freely, and remains indefinitely a gill-breather ; but if the water dries up it changes into the lung-breathing amblystoma.

---

* Boston Society of Natural History — anniversary memoir, 1880 Also, " American Naturalist," June, 1882.

† " Archives des Sciences," vol. liv, 1875.

We do not give this as examples of change of species, for the change is in the individual life, and therefore in the nature of metamorphosis, but as evidence of the power of physical conditions in modifying the development of organic forms and therefore of the manner in which gill-breathers were probably transformed into air-breathers.

To sum up : 1. All *inorganic* forms, without exception, have become what we find them by a natural process—i. e., by evolution. 2. All *organic* or living forms within the *limits of observation,* i. e., every living thing, has become what we now see, by a gradual, natural process—i. e., by evolution. 3. All taxonomic groups, except species, have undoubtedly become what we now see them by a gradual process, following the laws of evolution, and therefore presumably by a natural process of evolution. 4. By artificial means, breeds, races, etc., very similar, at least in many respects, to species, are seen to arise by a gradual natural process—i. e., by evolution. 5. In some instances, at least, natural species are observed to pass into one another by intermediate links in such wise that we are forced to conclude that they have been formed by a natural process.

May we not, then, safely generalize, and make the law universal ? Is not this a sufficient ground for confident induction ? Even though some facts are still inexplicable, is that a sufficient reason for withholding assent to a theory which explains so much ? In all induction we first establish a law provisionally from the observation of a comparatively few facts, and then extend it over a multitude of facts not included in the original induction. If it explains these also, the law is verified. The law of gravitation was first based on the observation of a few facts, and then verified by its explanation of nearly all the facts of celestial motion. There are some outstanding facts of celestial motion still unexplained, but we do not, therefore, doubt the law of gravitation. The same principle applied in biology ought to establish the law of evolution, for it also explains all the facts of biology as no other law can. But inductive evidence differs from other kinds of evidence in one respect, which, in fact, constitutes its strength to the scientific, but its weakness to the popular mind. It is a kind of circumstantial evidence, but its force does not consist in a few strong circumstances easily appreciated, such as strike the popular mind, and force conviction, but rather in a multitude of small circumstances, each by itself insignificant, but all together pointing to one conclusion and demanding one explanation. Such evidence is, indeed, overwhelming, but only to the mind that masters it. The evidence for the law of gravitation is literally the whole science of astronomy. So also the evidence for the law of evolution is the whole science of biology. Neither of these laws can be proved in a debating society, but only by a course of study. In the one case the law has been universally accepted—not, however, on evidence, for there are few indeed who appreciate the evidence, but on the authority of scientific unanimity. In the other case there has not yet been time enough for the already established unanimity to have its full effect.

Thus much, we believe, will be generally admitted as a very moderate claim. Evolution is certainly a legitimate induction from the facts of biology. But we are prepared to go much further. We are confident that evolution is *absolutely certain*. Not, indeed, evolution as a special theory—Lamarckian, Darwinian, Spencerian—for these are all more or less successful modes of explaining evolution ; nor evolution as a school of thought, with its following of disciples—for in this sense it is still in the field of discussion—but evolution as a law of derivation of forms from previous forms ; evolution as a law of continuity, as a universal law of becoming. In this sense it is not only certain, it is axiomatic. It is only necessary to conceive it clearly, to see that it is a necessary truth. This may seem paradoxical to some. I stop to justify it.

Physical phenomena we all admit follow one another in unbroken succession, each derived from a preceding, and giving origin to a succeeding. We call this the law of causation, and say that it is axiomatic. We might call it a law of derivation. So also organic *forms* follow one another in continuous chain, each derived from a preceding and giving origin to a succeeding. We call this a law of derivation. We might call it a *law of causation*, and say that it too is axiomatic. The origins of new phenomena are often obscure, even inexplicable, but we never think to doubt that they have a natural cause ; for so to doubt is to doubt the validity of reason, and the rational constitution of Nature. So also the origins of new organic *forms* may be obscure or even inexplicable, but we ought not on that account to doubt that they had a natural cause, and came by a natural process ; for so to doubt is also to doubt the validity of reason, and the rational constitution of organic Nature. The law of evolution is naught else than the scientific or, indeed, the rational mode of thinking about the origin of things in every department of Nature. In a word, it is naught else than the law of necessary causation applied to *forms* instead of phenomena. Evolution, therefore, is no longer a school of thought. The words *evolutionism* and *evolutionist* ought not any longer to be used, any more than *gravitationism* and *gravitationist* ; for the law of evolution is as certain as the law of gravitation. Nay, it is far more certain. The nexus between *successive events in time* (causation) is far more certain than the nexus between *coexistent objects in space* (gravitation). The former *is a necessary truth*, the latter is usually classed as a contingent truth. I have used and may continue to use the term evolutionist, but if so it is only in deference to the views of many intelligent persons, who do not yet see the certainty of the law.

CHAPTER II.

Introductory.

IT will be seen from the preceding chapter that we regard the law of evolution in its wider sense, viz., the derivative origin of all forms, organic or other, as axiomatic, and therefore requiring no further proof. Among scientific men there is no longer any discussion of the truth of this law, but only of the theories of the causes of the law. We believe that to the scientific mind there is no other rational mode of looking at the subject of origin of organic forms. To such a mind, therefore, all that follows is but the deductive application of that law in the explanation of the phenomena of organic Nature. But it takes time for the popular mind to readjust itself to new and revolutionary truth. Many minds, even among the most intelligent, have not yet accepted this as the only rational mode of thought. Many men require further *special proofs* of the derivative origin of organic forms. Even to those who accept evolution, these proofs will be interesting as illustrations of such origin. We will attempt to bring out these proofs under several heads, the most important of which are : 1. Proofs from morphology, or the general laws of animal structure ; 2. Proofs from embryology ; 3. Proofs from geographical distribution of organic forms ; and, 4. Proofs from artificial breeding. The subject is so vast that all we can do is to touch lightly only the most salient points under each of these heads ; for, as we have already said, the evidence is really nothing less than the whole science of biology. Preparatory to this, however, it is necessary to bring out a little more fully than before (page 29), though still only in outline, the two antagonistic views, which may be called the old and the new, or the natural and the supernatural, of the origin of new organic forms, especially species.

Origin of New Organic Forms; the Old View briefly stated.—According to the old-school naturalists, species are the ultimate elements of taxonomy : genera, families, orders, etc., may gradually change their character from age to age, by the introduction of new species ; but species were supposed to be substantially *permanent*. It was necessary to have some unit for convenience of description and classification, and this was found to be the best because most stable. As in nearly all cases of beliefs, this doctrine was held at first somewhat loosely, as a provisional and convenient view—as a good working hypothesis—but gradually, under pressure of controversy, became more strictly formulated, and, as it were, hardened into a scientific dogma, especially in the hands of

Agassiz. According to this view, the first pair or pairs of each specific kind originated we know not how, but certainly *at once in its present form* in full perfection, and, therefore, presumably by *direct creative* act of Deity ; and then afterward by the law of generation continued to produce others of the same pattern indefinitely. Moreover, the first one or more pairs of each kind multiplied and spread abroad in every direction, *each from its own center of origin*, as far as physical conditions and struggle for life with other species would allow. This idea explains tolerably well the geographical distribution of species as we now find it. For example, species on different continents are widely different, because those on each have originated independently where we now find them, and spread in all directions as far as physical conditions would allow, but could not reach other continents because of the ocean-barrier. That this is the only reason they are not there, is shown by the fact that, if they are carried there, they usually do perfectly well. Even on the same continent, for the same reason, species may be very different if separated by impassable barriers such as high mountain-chains or by climate. But wherever one group of species, originating in one place, comes in contact on the margin of their range with another group of species originating in another place, we see no evidence of *transmutation* of one form *into* another, but only *substitution* of one fully-formed species *for* another equally fully formed. Therefore, we must conclude that physical conditions may limit the range of a species, but can not transmute it into another. Thus, to say the least, many of the facts of geographical distribution are well explained by this idea of creative origin in specific centers and subsequent permanence of specific form. We say *many* of the facts ; we will show hereafter that *not all* can be thus explained.

But the main question is not of geographical but of geological distribution ; not distribution in space, but succession in time. Species do not continue forever. On the contrary, they have changed many times in the course of geological history. As conditions become unfavorable, species die out or become extinct, and others take their place and carry forward the life and development of the organic kingdom. Now, how do they change ? According to this school of thought, here also, as in geographical distribution, they are not transmuted but replaced ; here also physical conditions may destroy a species, but can not transform it into another. As species die out, others are created at once, out of hand and fully formed in their place ; but in accordance with a preordained plan consistently carried out and working ever toward higher and higher conditions. Thus, life is continued on the earth by the alternation of supernatural and natural processes ; by the alternate use of direct and indirect action of Deity : direct in the introduction of first pairs, indirect through the natural process of reproduction in the continuance and multiplication of the species. Each species is made according to a pattern in the Divine mind, on a sort of intellectual die, and then continues to reproduce a succession of individuals of the same pattern as if struck from the same die until the die is broken or worn out. Another die is made, of another pattern, and individuals are struck from this ; and so on, throughout the whole geological history of the organic kingdom. Only, we must add that the successive dies are made to follow one another according to a plan which is expressed by the three laws already given on page 11. Thus, the origin of individuals is natural, the origin of species supernatural ; the making of dies is supernatural, the coinage is natural.

We have stated this view in a too extreme form, in order to make it clearer. We now, therefore, proceed to qualify somewhat. Specific types were held, by writers of this school of thought, to be *substantially* but not absolutely unchangeable. Successive individuals of the same species were admitted to be not exactly alike. Such slight differences were called *varieties*. It was admitted, indeed, that species varied, but it was believed that such variations in any direction were strictly limited in amount. A species may be compared to a right cylinder standing on end. As such a cylinder may be tilted slightly in one direction or another, without overthrowing its equilibrium, the cylinder tending ever to right itself and return to its original position, so a species may be varied slightly in one direction or another without destroying its integrity, the species tending ever to return to its normal or typical form. But as the cylinder, if pushed too far from its normal position, is overthrown, so also a species, if pressed too far in the way of variation from its typical form, is destroyed, but not changed into another species. As cylinders may be more or less rigid, depending upon the breadth of their bases, so also some species are more rigidly set in their typical form, and some are more plastic to influences causing variations, but in all cases there is a limit to the amount of oscillation consistent with integrity.

The New View briefly stated.—According to Darwin, and all biologists of the present day, species are variable *without limit,* if only the causes of change are constant and slow enough in their operation, and the time long enough. A species must be in harmony with its environment, for this is the condition of its existence. Now, if the environment change, the species must *tend* to change slowly from generation to generation, so as to readjust its relations in harmony with the changing environment. If the change of environment be slow, the readjustment may be successful, and the species will change gradually into another form, so different that it will be called a different species, especially if the intermediate gradations be destroyed. If the change in the environment be too rapid, many species, especially the more rigid, will be destroyed, while the more plastic may survive by modification. Thus, at every step in the evolution of the organic kingdom, some species have died without issue, while others have saved themselves by changing into new forms in harmony with the new environment. Comparing to a growing tree, some branches overshadowed die, while others push on for light, forming new lateral buds, and dividing as they grow. By continued divergent change species gradually become genera, genera families, etc. Thus, varieties, species, genera, families, orders, classes, etc., are only different degrees of differences formed all in the same way. Varieties are only commencing species, species commencing genera, and so on. There is no making and wearing out of dies, and making of new ones; the whole process is a natural one—the whole series is genetically connected. In a perfect classification varieties, species, genera, families, orders, classes, etc., are only different *degrees of blood-kinship*.

So much may be regarded as certain, and out of the field of discussion among biologists of the present day. It is only in defining this process more accurately, and especially in the *theory of the causes* or *factors* of evolution, that there are still difference and discussion. The most probable view on this subject we now proceed to give.

Factors of Evolution.—The causes of change or adaptive modification, or the factors of evolution, are at least *four* well known, and probably many more still unknown : 1. The physical environment—heat and cold, dryness and moisture—affects function of organs, and function affects structure, and both changed function and changed structure are inherited by offspring, and so increased from generation to generation, becoming greater without limit. 2. Increased *use* or *disuse* of organs enforced or permitted by change in the environment, physical or organic, or both, induces change in form, size, and structure of the organs ; and this change is inherited by the offspring, and so from generation to generation small differences are integrated until they become great without limit. These two factors were recognized by Lamarck. 3. "Natural selection," or "survival of the fittest," among divergent varieties of offspring. This is the distinctive Darwinian factor. In the two preceding factors the change is during the *individual lifetime*, and reproduction is supposed to transmit it unchanged to the offspring. In this factor, on the contrary, the form and structure are supposed to remain unchanged during the individual life, but for some unknown cause there are slight variations in different directions (divergent) in the offspring from the same parents. Now, when we remember that by reproduction the number of individuals tends to increase by geometrical progression, and that in each generation only a very few (on an average only two from all the offspring of one pair) can survive, it is evident that among these divergent varieties those will most likely survive which are most in harmony with the external environment, and which possess the most efficient organs of defense or of escape, or for food-taking. The surviving offspring, therefore, will be on the average better in these respects than their parents. It matters not how little better, for the integration of even infinitesimal improvements from generation to generation will eventually produce any required amount of change. 4. To the above Darwin has added also *"sexual selection."* In *natural* selection there is struggle of *all* for *food*, or *means of living.* In sexual selection there is a struggle among the *males* for possession of the *female*, and the *means of procreation.* The one is connected with the nutritive appetite, the other with the reproductive appetite. This mode of selection acts in two ways, by the law of battle and the law of attractiveness. The strongest or the most attractive males alone, or mainly, leave offspring, which, of course, inherit their peculiarities; and these are increased indefinitely by integration through successive generations, thus increasing the strength or the beauty. Of these two laws, the law of battle is most conspicuous among mammals, and the law of attractiveness among birds. It is evident that this factor can not operate among many lower animals which are hermaphroditic, nor among plants.

Of these acknowledged factors of evolution, the first two were known to Lamarck and the older evolutionists. The third and fourth are distinctively Darwinian. According to Darwin, while all these are operative, the third is the most powerful; but Spencer accords this distinction to the Lamarckian factors. Many American zoölogists take the same view.

Such until very recently were all the recognized factors of evolution. But, within the past year (1886) has taken place, it seems to us, the most important advance in the theory of evolution since Darwin. It is the suggestion by Mr. Catchpool,* and afterward the more full elab-

---

* "Nature," vol. xxxi, p. 4, 1884.

oration by Dr. Romanes, of another factor, which he calls *"physiological selection."*

The great objections to the sufficiency of the theory of evolution, as left by Darwin, were twofold : 1. While natural selection accounts completely for the formation of *useful* structures or adaptive modifications, and therefore for differences characterizing classes, orders, families, and even genera—for these are all adaptive— it can not so completely account for those constituting species ; for these consist mostly of *trivial* differences in coloration, relative proportion of parts, which are of *no perceivable use* in the struggle for life, and therefore could not be preserved and integrated by natural selection. Therefore, according to Romanes, natural selection is a theory of origin of adaptive structures rather than of origin of species. Comparing to a growing tree, once admit lateral buds started, and natural selection completely accounts for the growth in different directions, and therefore for the profuse ramification ; but the origin of the lateral buds is not explained.

2. The second difficulty is as follows : Such commencing differences as constitute varieties and species not only would not be preserved and integrated by natural selection unless useful, but would immediately be *swamped by cross-breeding* with the parental form. But, as the whole divergence commences in varieties, evi-

---

* See abstract of Dr. Romanes's views, " Nature," vol. xxxiv, pp. 314, 336, 362. Also, discussions of the same by Meldola, Galton, Wallace, etc., in immediately subsequent numbers.

dently it could not commence at all unless this cross-breeding be in some way prevented.   This may, indeed, be done, without the assumption of any new factor of evolution, by *migration;* and, hence, migration must be regarded as an important agent in the creation of new forms, not only by the effect of a new environment, but also by prevention of the swamping of commencing species by cross-breeding with the parental form ; but in a crowded locality, without outlet for migration (the very conditions most favorable for severe competitive struggle, and therefore for most potent operation of natural selection ; and therefore, also, according to Darwin, for profuse diversification), commencing varieties could not pass into species, because swamped by cross-breeding. Once the divergence reaches the point of cross-sterility —i. e., of species—then, indeed, by true breeding, characters, even though not useful, may be preserved.   But how is it to commence ?

This difficulty has been severely felt by all Darwinists.   It seems to us that it is largely met by Dr. Romanes.   According to Romanes, no organ is so subject to varietal changes as the *reproductive,* and these in no respect so much as in degrees of fertility.   Unfortunately, these changes are not visible, and must be judged of only by the results.   It is not uncommon, for example, to find sterility between individuals (sexual incompatibility) who are both of them perfectly fertile with other individuals.   Similarly, cross-sterility, partial or complete, is not uncommon between varieties or races, as Mr. Darwin has long ago noticed. It very generally, as we know, occurs between, and, in fact, is constantly used as a test of, species. Now, this cross-sterility with parent stock, which we find so constant a character of species, and which, therefore, must *have commenced as a partial cross-sterility* in varieties, is it *antecedent or consequent to other variations?* It has been usual to suppose it consequent to a certain amount of divergence, viz., that which constitutes, or at least approaches, species. But, according to Romanes, it is *antecedent.* Among many other variations, this is that one which originates species, because it prevents reversion by cross - breeding with the parent stock, and insures true breeding with its own kind. In a word, it sexually isolates the species. Suppose, then, a species multiplying indefinitely in one locality : trivial variations of many kinds, and in many directions, occur among the offspring. These are merged by cross-breeding into the original type, which, therefore, remains unchanged. But, from time to time, among these variations there occur some affecting the reproductive organs in such wise as to produce partial or complete cross-sterility with the parent form. This is the beginning of a new species. It breeds true with its own kind, and therefore all the associated variations external and visible, and therefore constituting species, although trivial and of no use in the struggle for life, are preserved.

This view completely accounts for the cross-fertility of artificial breeds equivalent in other respects to species ; for cross-sterility is not an end aimed at by the breeder, it being easy to prevent cross-breeding, if desired, by artificial isolation. But, if this view be true, species from widely-different geographical regions ought also to be often cross-fertile, because, having been formed by geographical isolation, sexual isolation was not a necessary factor in their formation. This point deserves testing by careful observation.

It may be, and has been, objected to Dr. Romanes's claims, that this is no new factor; that physiological selection is only a form of natural selection. This objec‑ tion, it seems to us, is little more than a play upon words. It certainly is selection, and by a *natural* process, and therefore in some sense a natural selection, but not in the sense of Darwin. It is not a selection of individuals *fittest to survive ;* for cross-fertile individuals are as fit to survive as individuals, though not as species, as are cross-sterile. Natural selection is intent only on preserving the best individuals ; physiological selection on preserving the kind. Natural selection continues the direction of progress unchanged ; physiological makes new directions.

In addition to all these factors of *organic* evolution, there is still another far higher factor characteristic of man alone. This is the *conscious, voluntary co-operation of the thing evolving—the spirit of man—in the work of its own evolution.* This may be called the *rational factor.* This, the most important factor of human evolution, is usually ignored by writers on evolution—either as nonexistent, or else as lying beyond the domain of science.

We will emphasize its importance by taking it up more fully in the next chapter.

It will be observed that Darwin and his followers take divergent variations of offspring simply as a known fact, upon which natural selection operates to produce progressive modification ; and, as the cause of variation in offspring is wholly unknown, such variations are often spoken of as fortuitous. But, of course, it is well understood that nothing in Nature is really fortuitous. They may, however, for all purposes of natural selection be thus regarded until we know their cause. It is evident, then, that if we, with Darwin, take natural selection, as the most important known factor, the really most important cause of evolution is the *cause* of varieties. This is the *unknown* fundamental factor. As Darwin reduced Agassiz's three formal laws of succession to more general laws of life, and thus made one important step in the advance of biological science, so he who shall explain the *cause* of divergent variation will make another important step by reducing the phenomena to still more general and fundamental laws of life.

In conclusion, let me again impress upon the reader that all the doubt and discussion, above described, as to the factors of evolution, is entirely aside from the truth of evolution itself, concerning which there is no difference of opinion among thinkers.

CHAPTER III.

WE have given in the previous chapter six factors of
evolution—viz. : 1. *Pressure of the environment.* 2. *Use
and disuse of parts.* 3. *Natural selection.* 4. *Sexual
selection.* 5. *Physiological selection.* 6. *Reason.* Let
us now compare these as to their grade in the scale of
energy and as to the order of their introduction.

The first two or the Lamarckian factors are the lowest in position, the most fundamental and universal, and
therefore the first in the order of appearance. They precede all other factors, and were doubtless for a long time
the only ones in operation. For, observe, all the selective
factors—i. e., those of Darwin and Romanes—are conditioned on reproduction ; for the changes produced by these
are not in the individual during life, but in the offspring
at birth. And not only so, but the operations of these factors are further conditioned on *sexual modes* of reproduction ; for all the non-sexual modes of reproduction—as,
for example, by fissure and by budding—are but slight
modifications of growth, and the resulting multitude of

organisms may be regarded as in some sense *only an ex-tension of the first individual.* Of course, therefore, the identical characters of the first individual are continued indefinitely, except in so far as they are modified in successive generations by the effect of the environment and by use and disuse—i. e., by the Lamarckian factors. In sexual generation, on the contrary, the characters of two diverse individuals are funded in a common offspring; and the same continuing through successive generations, it is evident that the inheritance in each individual offspring is infinitely multiple. Now, the *tendency* to *variation* in offspring *is in proportion to the multiplicity of the inheritance:* for among the infinite number of slightly differing characters, as it were, offered for inheritance in each generation, some individuals will inherit more of one and some more of another character. In a word, sexual reproduction by multiple inheritance *tends to variation of offspring, and thus furnishes material for natural selection.**

Thus, then, I repeat, all the selective factors are absolutely dependent on sexual modes of reproduction. But there was a time when this mode of reproduction did not yet exist.† The sexual modes developed out of non-sexual modes. If these non-sexual preceded sexual modes of reproduction, it is evident that at first only Lamarckian factors could operate. Evolution was then

---

* This subject is more fully treated in chapter IX, p. 240 *et seq.*

† See an article entitled "Genesis of Sex," "Popular Science Monthly," 1879, vol. xvi, p. 167.

carried forward wholly by changes in the individual produced by environment and by use and disuse (acquired characters), inherited and increased by integration through successive generations indefinitely. It is probable, therefore, that the *rate* of evolution was at first comparatively slow; unless, indeed, as seems probable, the *earliest* forms *were then* and the *lowest* forms *are now* more plastic under the influence of physical conditions than are the present higher forms. Doubtless, now, in the higher animals and plants, the Darwinian factors are by far the most potent; for, among plants, where we can use these factors separately, if we wish to *make* varieties, we propagate by seeds (sexual reproduction); but, if we wish to preserve varieties, we propagate by buds and cuttings (non-sexual reproduction).

I have taken the two Lamarckian factors together, and showed that they preceded the Darwinian. But even in the two Lamarckian factors there is a difference in grade. Undoubtedly the lowest, the most fundamental, and therefore the first introduced, was *pressure of the physical environment*. For use and disuse of organs implies some degree of volition and voluntary motion, and therefore already some advance in the scale of evolution.

With the introduction of sex another entirely different and higher factor was introduced, viz., *natural selection*, or selection of the fittest individuals of a varying progeny. We have already seen how sexual generation produces variation of offspring, and how this furnishes materials for natural selection. As soon, therefore, as this form of generation was evolved, this higher factor came into operation and immediately assumed control; while the previous factors became subordinate, though still underlying, conditioning, and modifying the activity of the higher. The result was an immediate increase in the rate of evolution. It is very worthy of note that it is in the higher animals, such as birds and mammals, in which we have only the highest forms of sexual reproduction, where the diversity of characters of the two sexes funded in the offspring is the greatest, and where, therefore, the variation in offspring is also greatest and natural selection most active; it is precisely among these that the Lamarckian factors are most feeble, because, during the most plastic period of life, the offspring is removed from the influence of the physical environment, and from use and disuse by its inclosure within the womb, or within a large egg surrounded with abundant nutriment. Development is already well advanced before Lamarckian factors can operate at all.

Next, I suppose, physiological selection, or Romanes's factor, came into operation. After the introduction of sex, it became necessary that the individuals of some varieties should be isolated in some way, so as to prevent the swamping of varietal characters, as fast as formed, in a common stock, by *cross-breeding*. In very low forms, with slow locomotion, such isolation might easily take place accidentally. Even in higher forms, changes in physical geography or accidental dispersion by winds and currents would often produce geographical isolation, and thus, by preventing crossing with the parent stock, secure the formation of new species from such isolated varieties. But, in order to insure in all cases the preservation of commencing species, *sexual isolation*, or partial or complete infertility of some varieties with other varieties and with the parent stock, was introduced, as I suppose, later. The process by which this takes place has already been explained. According to Romanes, natural selection alone, with cross-breeding, tends to *monotypal* evolution; isolation of some kind is necessary for polytypal evolution. The tree of evolution, under the influence of natural selection alone, grows, palm-like, from its *terminal bud ;* isolation of varieties was necessary for the starting of *lateral buds*, and thus for the profuse ramification which is its most conspicuous character.

Next, I suppose, was introduced *sexual selection*, or contest among the males, by battle or by display, for possession of the females, and the success of the strongest or the most attractive; and the perpetuation and increase of these superior qualities of strength and beauty in the next generation. This, I suppose, was later, because connected with a higher development of the psychical nature. This is especially true where splendor of color or beauty of song determines the selection. As might be supposed, therefore, this factor is operative only among the highest animals, especially birds and mammals.*

* Mr. Wallace has recently, in his work on "Darwinism," taken strong ground against this Darwinian factor. He thinks, for example,

Next and last, and only with the appearance of *Man,* another entirely different and far higher factor was introduced, viz., *conscious, voluntary co-operation* in the work of his own evolution—a conscious, voluntary striving to *attain an ideal.* We have called this a factor, but it is much more than a mere factor, co-ordinate with other factors. It is, rather, a different kind of evolution. It is evolution on a higher plane and by another nature. As *physical* Nature works *unconsciously,* using certain factors, so *spiritual* nature works *consciously,* co-operating and using the same factors. At first this factor, if we still call it so, was extremely feeble. In the early stages of his progress, man, like other animals, was largely urged on by forces of organic evolution, unknowing and uncaring whither he tended. But more and more, as civilization advances, this higher and distinctively human factor becomes more and more dominant, until now, in civilized communities, it takes control of evolution. Reason, instead of Nature, now assumes control, though still using the methods and factors of Nature. This *free,* self-determined evolution of the race, in order to distinguish it from the *necessary* evolution of the organic kingdom, we call progress.

Now, in this whole process we observe two striking

---

that sexual vigor is the cause of both the splendor of color and the pertinacity which secures the female. We see little difference in this way of putting it. Our object, however, is not to argue the question of what are true factors, but simply to give the most accepted, and, as it seems to us, also the most probable view.

stages. The one is the introduction of sex, the other is the introduction of reason.* They may be compared to two equally striking stages in the development of the *individual.* As the *ontogenic* evolution receives fresh impulse at the moment of fertilization, so the evolution of the organic kingdom receives fresh impulse at the moment of introduction of sex. As in ontogenic evolution the individual at birth enters upon a new and higher plane, in which it co-operates in its own *physical* growth, so the organic kingdom, with the introduction of man, enters upon a new and higher plane, in which man co-operates in the physical and *spiritual* growth of the race. With sex three new and higher factors were introduced, and these immediately assumed control and quickened the rate of evolution. With reason another and infinitely higher factor is introduced, which, in its turn, assumes control, and not only again quickens the rate, but elevates the whole plane of evolution. Moreover, this voluntary, rational factor not only takes control itself, but transforms all other factors and uses them in a new way and for its own higher purposes.

This last is by far the greatest change which has ever occurred in the history of evolution. In organic evolu-

---

* By *reason* I mean the faculty of dealing with the phenomena of the *inner world of consciousness and ideas.* Animals live in one world — the outer world of *sense ;* man in two — the outer world of sense, like animals, but also in an inner and higher world of *ideas.* All that is characteristic of man comes of this capacity of dealing with the inner world. In default of a better word I call it reason. If any one can suggest a better word, I will gladly adopt it.

tion Nature operates by necessary law without the conscious voluntary co-operation of the thing evolving. In human progress man voluntarily co-operates with Nature in the work of evolution, and even assumes to take the process mainly into his own hands. Organic evolution is by *necessary* law, human progress by *free* or at least by freer law. Organic evolution is by a *pushing* upward and onward from *below* and *behind*, human progress by a *drawing* upward and onward from above and in front by the attractive force of ideals. In a word, organic evolution is by the law of *force*, human evolution by the law of *love*.

It may be well to stop a moment and show briefly some of the differences between organic and human evolution—differences which are, of course, wholly the result of the introduction of this new factor :

1. In organic evolution " *the fittest* " are those most in harmony with the physical environment, and therefore they survive. In human evolution *the fittest* are those most in harmony with *the ideal*, and often, especially in the early stages, when the race is still largely under the dominion of organic factors, they do not survive, because not in harmony with the social environment. But, although the fittest individuals may indeed perish, the *ideal* survives in the race and will eventually triumph.

2. In organic evolution the weak, the sick, the helpless, the unfit in any way perish and *ought to perish*, because this is the most efficient way of strengthening the *blood* or *physical nature* of the species, and thus of carrying forward evolution. In human evolution the weak,

the helpless, the sick, the old, the unfit in any way are sustained and *ought to be sustained*, because sympathy, love, pity, strengthen the *spirit* or *moral nature* of the race. But let us remember that in this material world of ours and during this earthly life the spirit or moral nature is conditioned on the physical nature; and, therefore, in all our attempts to help the weak we must be careful to avoid poisoning the blood and weakening the physical vigor of the race by inheritance. This gravest of social problems, viz., How shall we obey the higher law of love and mutual help without weakening the *blood* of the race by inheritance and the spirit of the race by removing the necessity of self-help?—this problem, I believe, can and will be solved by a *rational education*, physical, mental, and moral. I only allude to this. It is too wide a field to follow up here.

3. In organic evolution the bodily *form* and *structure* must continually change in order to keep in harmony with the ever-changing environment. In other words, organic evolution is by continual change of species, genera, families, etc. There must be continual evolution of new forms by modification. In human evolution, on the contrary, and more and more as civilization advances, man modifies the environment so as to bring it into harmony with himself and his wants, and therefore there is no necessity of change of bodily form and structure or making of new species of man. Human evolution is not by modification of *form*—new species; but by modification of spirit—new planes of activity, *higher character*. And the

spirit is modified and character elevated, not by *pressure* of an *external physical environment*, but by the *attractive* force of an *internal spiritual ideal*.

4. The way of evolution toward the highest—i. e., from protozoan to man and from lowest man to the ideal, the divine man—is a very *straight and narrow way*, and few there be that find it. In the case of organic evolution it is so straight and so narrow that any divergence therefrom is fatal to upward movement toward man. Once get off the track, and it is *impossible* to get on again. No living form of animal is on its way *manward*, or can by any possibility develop into man. They are all gone out of the way. There is none going right; no, not one. The organic kingdom developing through all geological times may be compared to a tree whose trunk is deeply buried in the lowest strata, whose great limbs were separated in early geological times, whose secondary branches diverged in middle geological times, and whose extreme twiglets, and also its graceful foliage, its beautiful flowers, and luscious fruits, are the fauna and flora of the present day. But this tree of evolution is an *excurrent stem*, continuous through the clustering branches to the terminal shoot—man. Once leave the stem as a branch, and it is easy to continue growing in the direction chosen, but impossible to get back on the straight upward way to the highest. In human evolution, whether individual or racial, the same law holds, but with a difference. If individual or race gets off the straight, narrow way toward the highest —the divine ideal—it is hard, very hard to get back on

the track. Hard, I say, but *not* impossible, because man's conscious voluntary effort is the chief factor in his own evolution. By virtue of self-activity, through the use of reason and co-operation in the work of evolution, man alone of all created things is able to rectify an error of direction and return again to the deserted way.

5. In organic evolution, when a higher factor appears, it immediately assumes control, and previous lower factors sink into a subordinate position, though still underlying and conditioning the higher. But in human evolution, the higher rational factor, when it comes in with man, not only assumes control, but transforms all other factors and uses them in a new way and for its own higher purposes. In fact, as already said, it is much more than a mere factor. It determines a new kind of evolution—evolution on a new and higher plane, though, indeed, underlaid and conditioned by the laws of organic evolution. As *external physical* Nature uses many factors to carry forward organic evolution, so the *internal spiritual* nature, characteristic of man alone, uses these same factors in a new way to carry forward human evolution or progress. Thus, for example, one organic factor—the environment—is modified or even totally changed so as to effect suitably the human organism. This is *hygiene.* Again, use and disuse—another factor—is similarly transformed. The various organs of the body and faculties of the mind are deliberately used in such wise and degree (determined by reason) as to produce the highest efficiency of each part and the greatest strength and beauty of the whole.

8

This is *education*—physical, mental, moral. So also the selective factors are similarly transformed, and *natural* selection becomes *rational* selection. We all know how this method is applied to domestic animals and cultivated plants in the formation of useful or beautiful varieties. Why should it not be applied also to the improvement of our race in the selection of our mates in marriage, or in the selection of our teachers, our law-makers, our rulers? Alas! how little even yet does reason control our selection in these matters! How largely are we yet under the law of organic evolution!

Application of these principles to some questions of the day :

I. Evolution, as a law of derivation of organic forms from previous forms by descent with modifications, as already shown, is as certain as the law of gravitation. This question has passed beyond the realm of doubtful discussion ; but the causes, the factors, the details of the process of evolution are still under discussion. Both Darwin and Spencer, the two great founders of the theory of evolution in its modern form, acknowledge and insist on at least four factors, viz., the two Lamarckian and the two distinctively Darwinian. The only difference between them is in the relative importance of the two sets : Spencer regarding the former and Darwin the latter as the more potent. But in these latest times there has arisen a class of biologists, including some of highest rank, such as Wallace, Weismann, and Lankester, who out-Darwin Darwin himself in their exaltation of the most distinctive Darwinian factor, viz., natural selection. They try to show that natural selection is the sole and sufficient cause of evolution; that changes in the individual, whether as the effect of the environment or by use and disuse of organs, are not inherited at all; that Lamarck was wholly wrong; that Darwin (in connection with Wallace) was the sole founder of the true theory of evolution; and, finally, that Darwin himself was wrong only in making any terms whatever with Lamarck. This view has been called *Neo-Darwinism*.

Perhaps the reasons for this view have been most strongly put by Weismann, and are based partly on experiments, but mainly on his ingenious and now celebrated theory of the immortality of germ-plasm. The animal body consists of two kinds of cells wholly different in function—somatic cells and germ-cells, including in this last the sexual elements both male and female. Somatic cells are specially modified for the various functions of the body; germ-cells are wholly unmodified. The somatic cells are for the conservation of the *individual* life, the germ-cells for the conservation of the *species*. In the development of the egg the germ-cell multiplies itself into a cell-aggregate, and then most of the resulting multitude of cells are modified in various ways to form the tissues and organs of the body—somatic cells; but a few are reserved and put aside in an unmodified form in the sexual organs as germ-cells, to again produce ova which again divide into somatic and germ cells, and so on indefinitely. Now, according to Weismann, inheritance is only through *germ-cells*, while the environment affects only the *somatic cells*. Therefore changes produced by the environment can not be inherited. Sexual modes of generation were introduced for the purpose of producing variability in progeny, and thus furnishing material for natural selection, as this was the only means of evolutionary advance. Weismann made many experiments on animals, especially by mutilation, to show that somatic changes are not inherited.

A full discussion of this question would be unsuitable in a work like this. We will therefore content ourselves with making three brief remarks:

a. If the views presented in the early part of this chapter are true, then the Lamarckian factors must be true factors, *because there was a time when there were no others.* They were therefore necessary, at least to start the process, even if no longer necessary at present.

b. But if these factors were ever operative, *they must be so still*, though possibly in a subordinate degree. A lower factor is not abolished, but only becomes subordinate to a higher when the latter is introduced. Thus it may well be that Lamarckian factors are comparatively feeble at the present time and among living species, especially of the higher animals, and yet not absent altogether. In the earliest stages of evolution there was a *complete identification of germ-cells and somatic cells*—of the individual with the species. In such cases, of course, any effect of the environment must be inherited and in-

creased from generation to generation. But the differentiation of the germ and somatic cells was not all at once,
nor is their sympathetic relation completely severed. It
was a *gradual process*, and therefore the effect of the
environment *on the germ-cells through the somatic cells*
continued, though in decreasing degree, and still continues. The differentiation in the higher animals is now
so complete that germ-cells are probably not at all affected by changes in somatic cells, unless these changes
are *long continued in the same direction, and are not
antagonized by natural selection.*

c. It is a general principle of evolution that the *law
of the whole is repeated with modifications in the part.*
This is a necessary consequence of the unity of Nature.
We ought to expect, therefore, and do find, that the order
of the use of the factors of evolution is the same in the
evolution of the *organic kingdom*, in the evolution of
each species, and in the evolution of *each individual.* In
all these the physical factors are at first powerfully operative; these become subordinate to organic factors, and
these, in their turn, to psychical and rational factors.
Therefore, as the individual in its early stages—i. e., in
embryo and infancy—is peculiarly plastic under the influence of the physical environment, and afterward becomes
more and more independent of these; so a species when
first formed is more plastic under the influences of Lamarckian factors, and afterward becomes more rigid to
the same. And so also the organic kingdom was at first
more plastic under Lamarckian factors, and has become

less so in the present species, especially in the higher animals. The principal reason of this, as we have already seen, is the increasing differentiation of germ and somatic cells, and the removal of the former to the interior, where they are more and more protected from external influence.

II. Some evolutionists—the materialistic—insist on making human evolution identical in all respects with organic evolution. This, we have shown, is not true. The very least that can be said is that a new and far more potent factor is introduced with man, which modifies greatly the process. But we may claim much more, viz., that evolution is here on a different and higher plane. The factors of organic evolution are, indeed, still present, and condition the whole process; but they are not left to be used by Nature alone. On the contrary, they are used in a new way and for higher purposes—by reason.

But by a revulsion from the materialistic extreme some have gone to the opposite extreme. They would place human progress and organic evolution in violent antagonism, as if subject to entirely different and even opposite laws; but we have also shown that, although the distinctive human factor is indeed dominant, yet it is underlaid and conditioned by all the lower factors; that these lower factors are still necessary as the agents used by reason.

III. We have already given the views of Weismann and Wallace, and some reasons for not accepting them;

but there is one important aspect not yet touched. There are some logical consequences of these views when applied to human evolution which seem to us nothing less than a *reductio ad absurdum*. This brings into view still another contrast between organic evolution and human progress.

In organic evolution, when the struggle for life is fierce and pitiless as it is now among the higher animals, natural selection is undoubtedly by far the most potent factor. It is at least conceivable (though not probable) that at the present time organic evolution might be carried on mainly or even wholly by this factor alone; but in human evolution, especially in civilized communities, this is impossible. If Weismann and Wallace be right, then alas for all our hopes of race improvement—physical, mental, and moral!—for natural selection will never be applied by man to himself as it is by Nature to organisms. His spiritual nature forbids. Reason may freely use the Lamarckian factors of environment and of use and disuse, but is debarred the unscrupulous use of natural selection *as its only method*. As this is an important point, we must explain.

All enlightened schemes of physical culture and hygiene, although directed primarily to secure the strength, the health, and the happiness of the *present generation*, yet are sustained and ennobled by the conviction that the improvement of the individuals of each generation enters by inheritance into the gradual physical improvement of the race. All our schemes of education, intellectual and moral, though certainly intended mainly for the improvement of the individual, are glorified by the hope that the race also is thereby gradually elevated. It is true that these hopes are usually extravagant; it is true that the *whole* improvement of one generation is not carried over by inheritance into the next; it is true, therefore, that we can not by education raise a lower race up to the plane of a higher in a few generations or even in a few centuries : but there must be at least a small residuum, be it ever so small, carried forward from each generation to the next, which, accumulating from age to age, determines the slow evolution of the race Such are the hopes on which all noble efforts for race-improvement are founded. Are all these hopes baseless ? They are so if Weismann and Wallace are right. If it be true that reason must direct the course of human progress, and if it be true also that selection of the fittest in the organic sense is the only method which can be used by reason, then the dreadful law of pitiless destruction of the weak, the helpless, the sick, the old, must with Spartan firmness be voluntarily and deliberately carried out. Against such a course we instinctively revolt with horror, because contrary to the law of our spiritual nature.

But the use by reason of the Lamarckian factors is not attended with any such revolting consequences. All our hopes of race-improvement, therefore, are strictly conditioned on the efficacy of these factors—i. e., on the fact that useful changes, determined by education in each generation, are to some extent inherited and accumulated in the race.

CHAPTER IV.

General Principles.

Analogy and Homology.—In biology those organs or
parts in different animals are said to be *analogous* which,
however different their origin, have a general similarity
of form and especially of function ; while those are called
homologous which, however different their general appearance, and however different their function, yet may,
by close examination and extensive comparison, be shown
to be modifications of one another—to be, in fact, originally the same part modified for different purposes. In
the former the parts compared look and behave as if they
were the same, but are not ; in the latter they look and
behave entirely differently, but are, in fact, the same
part in disguise.

We can best make this plain by examples. The wing
of a bird and the wing of a butterfly are analogous organs. They have the same function—i. e., flying ; and
this function necessitates the same general form of a flat

plane. But they are not at all homologous ; they are not at all the same organ or part. They certainly have never been formed one out of the other by modification. But the wing of a bird, the fore-paw of a reptile or mammal, the wing of a bat, and the arm and hand of a man, though so different in form and function, are homologous parts. On close examination they are found to have the same general structure, to be composed of essentially the same pieces, although they are so greatly modified in order to adapt them to different functions, that the general or superficial resemblance is now lost. Their structure is precisely such as it would be if they had all originated from some archetypal fore-limb by modifications in different directions of its several parts. By extensive comparison in the taxonomic and ontogenic series, all the intermediate gradations between these extreme modifications may be picked up.

Another example. The lungs of a mammal and the gills of a fish are analogous organs, since they have the same function of aëration of the blood. But they are not at all homologous : they are not built on the same plan ; by no effort of the mind can we imagine that the former could have come out of the latter by modification. On the contrary, we have positive proof that it did not so come. But there is an organ in the fish which is homologous with the mammalian lung, viz., the air-bladder, or swim-bladder. We know it—1. Because we can trace in the taxonomic series all the gradations from the one to the other. In most fishes the air-bladder is

wholly cut off from the gullet, and only very feebly supplied with blood. It is used and can be used only for
flotation. In others, as the gar-pike, the swim-bladder
is quite vascular and opens by a tube into the throat.
Through this opening air is gulped down from time to
time into the bladder, and again from time to time ex-
pelled. In other words, this fish supplements its gill-
breathing by an imperfect lung-breathing. We have here
the beginning of a lung. In still other fishes, viz., the
Dipnoi (*lepidosiren* and *ceratodus*, Fig. 2), the air-blad-

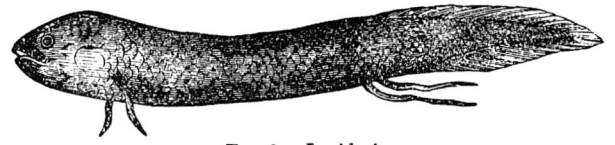

FIG. 2.—Lepidosiren.

der becomes a more perfect lung—i. e., a very vascular
sacculated sac ; and there is not only an opening into
the throat, but also from the throat to the snout. In
other words, we have for the first time *nostrils*. These
fishes completely combine gill-breathing with lung-
breathing. The step from these to the lowest amphibian reptiles is so small, that some have classed the
lepidosiren among amphibians instead of fishes. The
siredon or axolotl of New Mexico, the necturus or menobranchus of our Northern lakes, and the siren of our
Southern swamps, have both gills and lungs, and breathe
both air and water ; but the lung is very imperfect, being
only a sacculated sac, like the air-bladder of the cerato-

dus and lepidosiren. No one doubts that the air-breathing organ of an amphibian is a true lung ; yet we have
traced all the gradations between it and the air-bladder
of a fish. We conclude, therefore, that if there be any
such thing as transmutation of organic forms, the lung
of higher animals must have been formed by the process
above described.*

But we know it still more certainly—2. Because we
can trace the change from the one to the other in the ontogenic series. In the life-history of the individual we
can actually see the one thing change into the other. The
frog, as is well known, when first hatched, is a tadpole.
It has no legs, but locomotes by means of a vertically-
expanded tail. It has no lungs, but breathes water instead of air, by means of gills. It is in all respects,
therefore, a fish, and would be classed as such if it remained in this condition. But it does not ; it gradually
loses its tail and gills, and acquires legs and lungs, and
breathes air only. Now in this change whence came the
lungs ? From the gills by modification ? No ; but
from an organ similar in character and position to the
air-bladder of a ceratodus, or a lepidosiren. This organ
has gradually developed into a lung. The steps of the
change are briefly as follow : First, the breathing is
wholly water-breathing by gills. Next, by the development of this other organ, it is partly water-breathing by

---

* While all comparative anatomists agree that the lung is a diverticulum from the œsophagus, like the air-bladder of the gar-fish, some
think that it is a *different* diverticulum, which is seen first in the dipnoi.

gills, and partly air-breathing by lungs. Lastly, the gills gradually dry up, and the lungs develop more and more, until the breathing is wholly by lungs.

We have dwelt somewhat upon this example, because it is an excellent example of what we mean by homology, and also because we will have occasion to use it again. But so important, for all that follows in this part, is a clear idea on the subject of homology, that it will be best to familiarize the mind of the reader with it by means of a few examples drawn from plants.

A potato is analogous to a root—a tuberous root like that of a dahlia or a sweet-potato—but is not at all homologous with these. On the contrary, it is homologous with a stem. It is essentially an underground, leafless branch, which has thickened enormously at the point by accumulation of starch. The evidence of this is found in the fact that it has rudimentary leaves (scales) arranged in regular spiral order of phylotaxis, each with its axillary bud (eyes). It is still more clearly shown by the fact that buds above-ground which, if let alone, would form leafy branches, may be made to become tubers by covering them with earth or dead leaves, and thus excluding the light ; and, conversely, underground buds which, if let alone, would form tubers, may be made to grow into leafy branches by exposing them to the light.

Take another example : The broad, flat, elliptical, green masses so characteristic of the cactus family, and usually called their leaves, are indeed *analogous* to leaves in color, form, and function; for they are green and flat, and assimilate carbonic acid and water (CO_2 and H_2O) like leaves. But they are not, in truth, leaves, but modified stems, for they have the essential structure of stems, with their pith, wood, medullary rays, and bark, and may be traced through all gradations into the ordinary cylindrical form of stems. Where are their leaves, then? Their spines are their abortive leaves. These are arranged spirally like leaves, and bear buds in their axils like leaves. They are, in truth, leaves, modified to perform the function of defensive armor; while their function has been delegated to the stem flattened for this purpose.

One more example : The acacias, of which there are fifteen to twenty species in California, introduced from Australia, form two groups having extremely different styles of leaves. We will call them the feather-leaved and the simple-leaved acacias. In the former, the leaves are very finely bipinnate, and the general aspect of the foliage is extremely feathery and graceful. In the latter the leaves are simple, ovate, and, curiously enough, set on edge ; and the general aspect of the tree is therefore rather stiff. It seems at first incredible that leaves so different and aspects so diverse should belong to plants of the same genus. But a little close examination shows that, as usual, the botanists are right and the popular judgment wrong. The plumose-leaf is the normal leaf-form for this genus. The simple leaf is not only abnormal, but in a homological sense is not a leaf at all—i. e., it does not correspond to the part called the *blade* in ordinary simple leaves of other trees. In the seedling of the simple-leaved acacias, and sometimes for a considerable time in the young tree, the leaves are all plumose. As the tree matures it gradually changes its dress and puts on its *toga virilis*. The gradual change from the

FIG. 3.—A branch of young acacia, showing change from one form of leaf to the other; *a, b, c, d,* successive stages of change; *l, s,* leaf stalk which gradually changes into the blade in *c, d,* and *e.*

one form to the other may easily be traced in the same tree, and even often in the same branch (Fig. 3). The steps of the change (*a*, *b*, *c*, and *d*) are shown in the following figure, drawn from nature. It is seen, by bare inspection of the figure, that the so-called leaf, *d*, of the simple-leaved acacias, is really the vertically-expanded leaf-stalk, *l*, *s*, the true leaf or blade being wholly aborted. The whole structure of this so-called leaf is different from that of a true blade. For example, its style of ribbing is parallel, its position is edgewise to the sky, its palisade cells are on both sides alike, etc. To emphasize this difference, botanists call such an apparent leaf a *phyllodium*, or phyllode.

After these illustrations we now repeat the definitions in different words. Analogy has reference to *general resemblance* of form determined by *similarity of function*, however different the origins of the parts compared may be. Homology has reference to *community of origin*, however obscured to the superficial observer such common origin may be by modifications necessary to adapt to different functions. Observe, then, there are two ideas here which must be kept distinct. One is common origin, always shown by deep-lying, essential identity of structure ; the other is adaptive modification for function. Organs of the most diverse origin may resemble by adaptive modification for the same function. This is analogy. Organs of the same origin may assume very different appearance by adaptive modifications for different functions. This is homology. In the latter case, which is the one that concerns us, a profound study of essential structure and structural relations to other parts, and especially extensive comparison in the taxonomic and ontogenic series, will usually detect the homology, or common origin, in spite of the obscurations produced by adaptive modifications. It is seen, also, that analogy is a superficial resemblance, easily detected by the popular eye, and therefore embodied in popular language ; while homology is a deep-seated and essential resemblance, detected often only by profound study and extensive comparison. Now, one of the strongest proofs of the truth of evolution is taken from the homologies of animal structure. Common origin completely explains homology. Every other explanation is transcendental, and therefore unscientific.

Primary Divisions of the Animal Kingdom.—Now, the animal kingdom consists of several primary divisions, called sub-kingdoms or departments. The animals in these groups differ so essentially from one another in their *plan of structure*, that it is difficult, if not impossible, to trace any structural relation between them—to imagine how the members of one could have been derived from those of another—or conceive the common stem from which they all separated. In other words, it is impossible, in the present state of knowledge, to trace homology with any certainty from one group to another. But within the limits of each primary group the homology is easy. Some naturalists—Agassiz and Cuvier—have made four or five of these primary groups. Some—Huxley—have made eight. Some make nine or

9

ten.* We will not trouble ourselves to settle this question ; for all agree to make *vertebrata* and *articulata* or *arthropoda* two of them, and all our illustrations will be drawn from these. Other groups are too unfamiliar to the general reader to serve our purpose.

Now, as already stated, homology can not be traced with any certainty between the primary groups, but within the limits of each group it may be traced with ease and beauty. Analogy, however, being connected with function, and function being universal, can be traced throughout the animal kingdom. While, therefore, it is probable, nay, almost certain, that all animals have had a common origin, we can not yet trace these great departments by homology to that common origin. But the common origin of each department is quite clear. For example, the structure of all vertebrate animals is precisely such as would be the case if all came from one primal vertebrate, variously modified to adapt to various modes of life. Also, the structure of all arthropods is precisely such as would be if all came from one primal arthropod, which, from generation to generation, became gradually modified in different directions, in order to adapt itself to various modes of life. But between

---

* Undoubtedly the true principle on which primary groups ought to be made is, *identity of general plan of structure*, or *traceableness of homology throughout.* For these groups are the great primary branches of the tree of life, and classification ought to represent degrees of genetic relationship. This was Agassiz's principle, although he did not admit the genetic relation. This principle has been, it seems to us, too much neglected by later systematists.

arthropods and vertebrates we can not yet clearly see a common origin, although there doubtless was such.

These great departments may, therefore, be compared to *natural styles of animal architecture.* As there are various styles of human architecture—Oriental, Egyptian, Greek, Gothic—each of which may be variously modified to adapt it to all the different purposes for which buildings are made, without destroying, though perhaps obscuring, the integrity of the style; so the different primary groups or departments may be regarded as different styles of animal structure, each of which may be and has been modified in many ways to adapt it to various habits and modes of life, obscuring but not destroying the general style. Or they may be compared to natural *machines.* As a steam-engine, by modification, may be adapted to many kinds of purposes, obscuring, perhaps, but not destroying the essential identity of structure; even so the vertebrate machine by modification may be, and has been, adapted to many kinds of purposes, and thus become a swimming-machine, a crawling-machine, a flying-machine, a running- and leaping-machine, without destroying, although obscuring, the essential identity of structure. As in architecture, æsthetic principles of form may be traced through each style, but not from style to style, while the mechanical principles of construction run through all alike; so also in animal architecture, the laws of form and styles of structure are traceable with ease only within the limits of each primary group, while the laws of function are traceable through all groups alike. Or, again, and finally : Each of these departments may be compared to a *tree*, with branches, twigs, and spray, all obviously coming from one common stem, but each stem seems separate. They are, indeed, probably, themselves only great branches of one common trunk, but their connection is too remote and obscure to be made out clearly by means of homology. Other evidences, however, drawn from other sources, as we shall see hereafter, are not wholly wanting.

CHAPTER V.

THE proposition to be established here is, that all vertebrates have not only a common general plan of structure, but an essential identity even in detail, although
this identity is obscured by adaptive modifications. We
will try to show first a common general plan, and then,
taking parts most familiar to the general reader, will
show essential identity even in detail.

Common General Plan.—1. All vertebrate animals, and
none other, have an *internal* jointed skeleton worked by
muscles on the *outside*. As we shall see hereafter, the
relation of skeleton and muscle in arthropods is exactly
the reverse.

2. In all vertebrates, and in none other, the axis of
this skeleton is a jointed backbone (vertebral column)
inclosing and protecting the nervous centers (cerebro-
spinal axis). These, therefore, may well be called back-
boned animals.

3. All vertebrates, and none other, have a number of
their anterior vertebral joints enlarged and consolidated

into a box to form the skull,* in order to inclose and protect a similar enlargement of the nervous center, viz., the brain ; and also usually, but not always, a number of posterior joints, enlarged and consolidated to form the pelvis, to serve as a firm support to the hind-limbs.

4. All vertebrates, and none other, have two cavities, inclosed and protected by the skeleton, viz., the neural cavity above, and the visceral or body cavity below, the vertebral column ; so that a cross-section of the body is diagrammatically represented by Fig. 4.

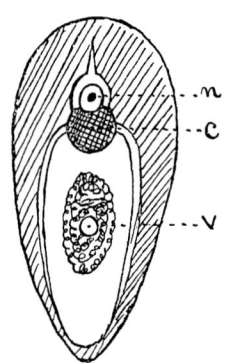

5. All vertebrates, with few exceptions, and no other animals, have two and only two pair of limbs. The exceptions are of two kinds, viz. : *a*, some lowest fishes, amphioxus and lampreys, which probably represent the vertebrate condition before limbs were acquired ; and *b*, degenerate forms like snakes and some lizards, which have lost their limbs by disuse.

FIG. 4.—Diagram cross-section through the body of a vertebrate, showing the relation of skeleton to the cavities. *n*, neutral cavity ; *v*, visceral cavity; *c*, centrum of vertebra.

So much concerns the general plan of skeletal structures, and is strongly suggestive of—in fact, is inexplicable without—common origin. But much more remains which is not only suggestive, but demonstrative of such origin. By extensive comparison in the taxonomic and

---

* The Amphioxus, the lowest of all vertebrates—if vertebrate it may be called—is an exception to 2 and 3. In this animal the vertebrate type is not yet fully declared.

ontogenic series, the whole vertebrate structure in all its details in different animals may be shown to be modifications one of another. Sometimes a piece is enlarged, sometimes diminished, or even becomes obsolete; sometimes several pieces are consolidated into one; but, in spite of all these obscurations, corresponding parts may usually be made out. This is the main subject of this chapter.

Special Homology of Vertebrate Limbs.—It would lead us much too far into unfamiliar technicalities to take up the whole skeleton. We select the limbs, both because their general structure is more familiar, and because in them the two fundamental ideas of essential identity and of adaptive modification are both admirably illustrated. The reason of this is, that it is by the limbs that the organism chiefly reacts on the environment, and is modified by it.

Fore-limbs.—In the accompanying figures (Figs. 5–18) we have represented, side by side, the fore-limbs of many vertebrates, taken from all the classes—mammals, birds, reptiles, and fishes. For convenience of comparison, the corresponding parts are similarly lettered in all. Also, in order to identify easily certain important corresponding segments, we have drawn through them a continuous dotted line. In man, nearly all the parts are present, and his limbs, therefore, may be taken as a term of comparison; for man's structure, except his brain, is far less modified than that of many animals.

Note, then, the following points : 1. The collar-bone (clavicle) is associated with wide separation of the shoulders, and the free use of the fore-limb for prehension or for flight, but is gradually lost in proportion as the fore-limb is brought nearer together and used for support,

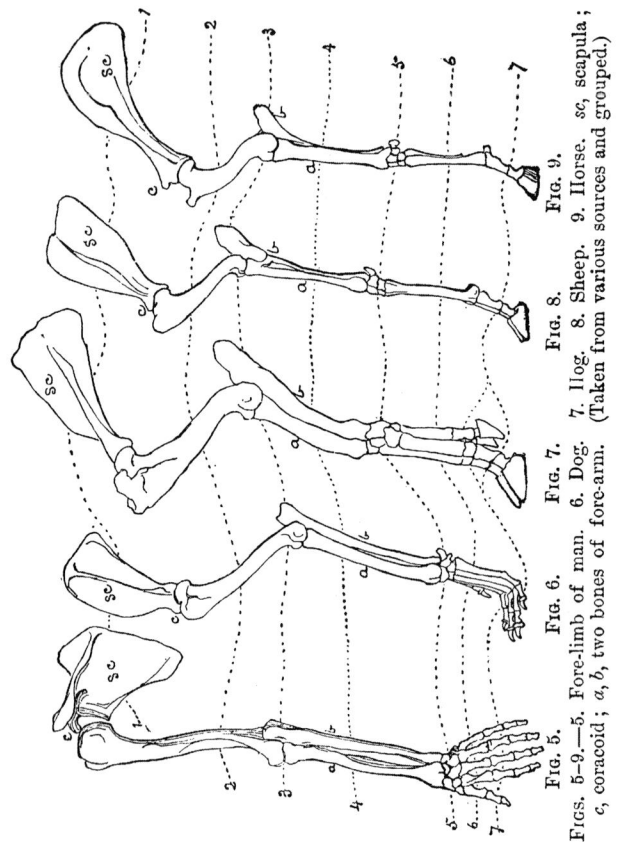

Figs. 5–9.—5, Fore-limb of man. 6, Dog. 7, Hog. 8, Sheep. 9, Horse. sc, scapula; c, coracoid; a, b, two bones of fore-arm. (Taken from various sources and grouped.)

because it is no longer wanted. I say *gradually*, for all the steps of the passing away may be found. The use-less rudimentary condition is not uncommon.

2. The coracoid (c), it is seen, is a small, beak-like process of the blade-bone (scapula) in man and mammals ; but in birds (Fig. 11) and reptiles (Figs. 14, 18)

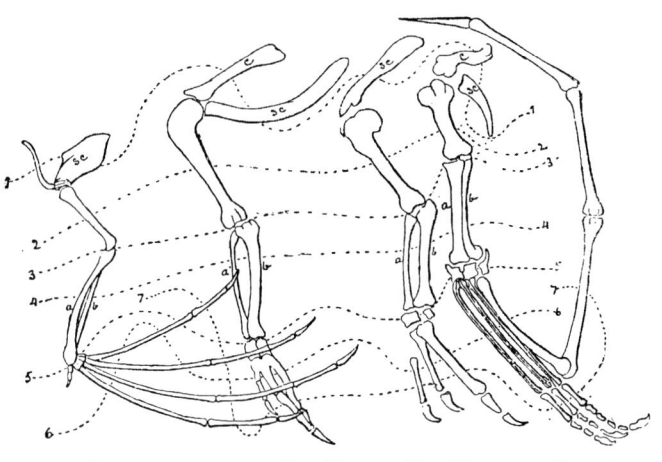

Fɪɢ. 10.   Fɪɢ. 11.   Fɪɢ. 12.   Fɪɢ. 13.
Fɪɢs. 10–13.—10. Fore-limb of bat.   11. Bird.   12. Archæopteryx.   13. Pterodactyl.   (Lettered as in previous figures ; grouped from various sources.)

it is a separate bone as large as the blade-bone itself, jointed with the latter at the shoulder and with the breast-bone (sternum) in front, thus making together a strong shoulder-girdle for the attachment of the fore-limb.   This was undoubtedly the condition in the original or earliest walking animal, viz., reptiles.   It was inherited and retained by birds, because necessary for powerful action of the wings in flight.   In mammals it gradually dwindled and became united with the blade-bone as a process.   In one mammal, the lowest and most reptilian living—the ornithorhynchus—the coracoid is much like that of reptiles—a large, flat bone, separated from the blade-bone and articulated with the breast-

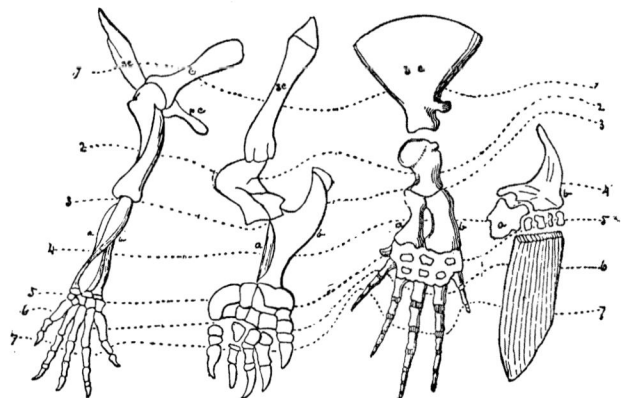

FIG. 14. FIG. 15. FIG. 16. FIG. 17.
FIGS. 14–17.—14. Fore-limb of turtle. 15. Mole. 16. Whale. 17. Fish.

bone. It is a significant fact that, in the mammalian embryo, it is first developed as a separate bone and afterward united with the scapula.

3. In man, monkeys, bears, and some other mammals, the limb is fairly free from the body and the elbow half-way down the limb ; while in herbivores (Figs. 8, 9), such as the horse, ox, and deer, etc., the elbow is high on the side of the body, and the limb is free only from the elbow downward. Perhaps in these cases most observers do not recognize it as an elbow at all. All gradations between these extremes are easily traced. The free condition of the limb is evidently the original one, the condition in herbivores being an extreme modi-

fication associated with another modification mentioned under 5.

4. In man and in many mammals, and in all reptiles and birds, there are two bones in the forearm (radius and ulna). In the more spe-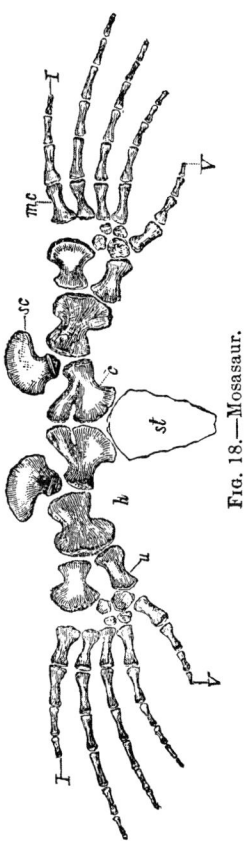cialized forms of hoofed animals (ungulates), such as horse and ruminants (Figs. 8, 9), there is apparently but one. Two is the normal and original number; but one of them, the ulna, has gradually become smaller and smaller, and finally is reduced to a short splint, and consolidated with the radius as a process extending backward to form the point of the elbow. In the horse family every step of this reduction and consolidation may be traced in the course of its geological history.

5. The *wrist* of many mammals and all birds differs in structure from that of man, chiefly in containing a smaller number of bones. The normal number, as in man, seems to be eight. The decrease takes place mainly by consolidation of two or more into one. In such

cases usually the embryo will show the bones still separate, thus revealing the ancestral condition. Again, the *position* of the wrist is noteworthy. In man, monkeys, the bear family, and several other mammalian families, and in all reptiles, the hand bends forward at the wrist, so that the tread is on the whole palm (palmigrade). But, in all the most specialized mammals, the wrist can not bend in this direction, and therefore this joint can not be brought to the ground. The tread is therefore on the toes (digitigrade), and the wrist is high up above the ground. In the horse (Fig. 9), the ox, and many other mammals, for example, the wrist is so high that it is not usually recognized as a wrist, and is often called the *fore-knee.* Now, homologous parts ought to have the same *scientific* name ; but to use the word *"hand"* in the case of lower animals might produce confusion and misconception. Therefore it has been agreed among comparative anatomists to use instead the Latin word *"manus"* for all that corresponds, in any animal, to the hand of man—i. e., all from the wrist downward. The manus of a horse is about fifteen inches long. The manus of a pterodactyl, such as that found by Marsh in the cretaceous strata of the West, with an expanse of wings of twenty-five feet, was probably not less than seven or eight feet long.

6. The number of palm-bones (metapodal) and toes deserves special notice. In fishes, and in some extinct swimming reptiles, these are or were very numerous, but in the earliest land-animals they became five. This is

the number now in nearly all reptiles, and in all the
more generalized mammals. It may be called the normal
number for a walking animal. In very many mammals,
such, for example, as the dog family, they are reduced to
four, though the fifth often remains as a useless, rudimentary splint and dew-claw (Fig. 6), thus showing the
process of dwindling in the ancestry. In hoofed animals the process of gradual diminution is shown even
in existing forms, and still better in extinct forms.
Confining ourselves, now, only to existing forms, in the
elephant there are five palm-bones and toes, and in the
hippopotamus there are four, all functional. In the hog
(Fig. 7) there are still four, but two are behind the
others and much smaller, and do not touch the ground
—are not functional unless in soft ground. In the cow,
deer, etc., the palm-bones are reduced to two, and these
are consolidated into *one* (canon-bone), and the toes are
reduced to two efficient and two useless rudiments. In
the sheep and the goat (Fig. 8) these useless rudiments
are dropped, and there are two only. Finally, in the
horse (Fig. 9), the *toes* are reduced to one, although the
palm-bones are still three, two of them, however, being
reduced to rudimentary splints.

How is it with birds ? Have these also palm-bones
and fingers ? Yes, in birds (Fig. 11) there are three
palm-bones and three fingers (the fourth and fifth being
wanting) ; one of them—the thumb—is free, and sometimes carries a claw. In the earliest known and most
reptilian bird, the archæopteryx (Fig. 12), all the three

fingers are free, have the full number of joints, and all of them carry claws. In the embryo of living birds the fingers are all free, as in the archæopteryx.

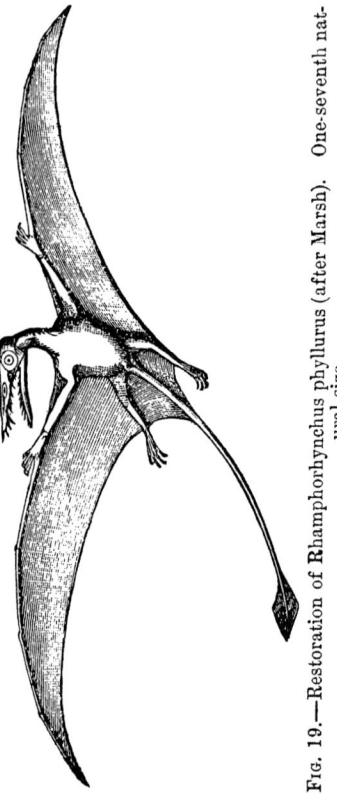

FIG. 19.—Restoration of Rhamphorhynchus phyllurus (after Marsh). One-seventh natural size.

7. Observe, finally, as an admirable illustration of different adaptative modifications for the same purpose — flight — the structure of the manus of flying animals. In the bat (Fig. 10), the flat flying-plane is made by enormous elongation of the palm-bones and finger-bones, their wide separation and the stretching of a thin membrane between them. In the pterosaurs, or extinct flying reptiles (Fig. 13), one finger only is greatly enlarged and elongated, and the flying-membrane is stretched between it and the hind-leg (Fig. 19), while the other three fingers are free and provided with claws. If it be asked which finger is it that is so greatly enlarged in this animal,

we answer, it is the *little finger*. In birds, on the contrary, the manus is consolidated to the last degree, to form a strong basis for attachments for the quills which form the flying-plane, and which are themselves extreme modifications of the scales of reptiles. But throughout all these extreme modifications the same essential structure is detectable.

It is perhaps unnecessary to dwell upon the still greater modifications of limbs for swimming, as in the whale (Fig. 16), the ichthyosaur, mosasaur (Fig. 18), and the fish (Fig. 17). A careful inspection of the figures, after what we have said, will be sufficient to explain them. In the fish alone the upper segments of the limb, viz., shoulder-girdle and humerus, are wanting, not being yet introduced, and the manus is not yet differentiated into palm-bones and fingers, and the fingers are indefinitely multiplied. All these characters are indications of low position in the scale of evolution. The earliest vertebrates were fishes. Limbs were not yet completely formed. In embryos of higher animals, also, the outer segments are first formed.

Hind-Limbs.—Figs. 20 to 24 represent, in a similar way, the hind-limbs of several animals—in this case all mammals. As before, corresponding parts are similarly lettered, and a dotted line is carried through certain prominent parts, especially the knee, heel, instep, and toes. By careful inspection the figures explain themselves. Nevertheless, it will be well to draw special attention to several of the more important points:

1. See, then, the position of the knee. The thigh-bone in man, monkeys, bears, and several other families of mammals, and all reptiles, is free from the body, and

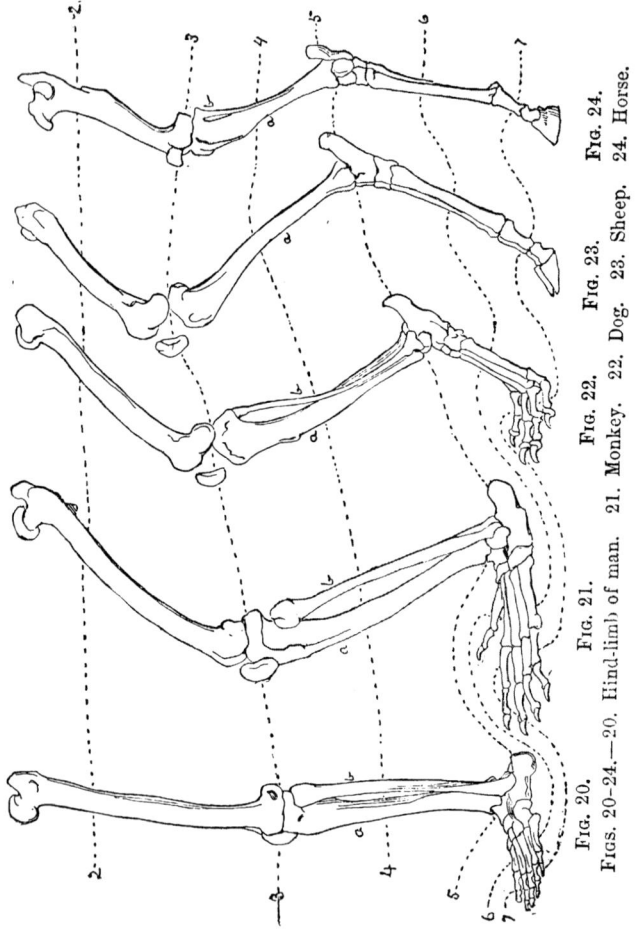

FIG. 20. FIG. 21. FIG. 22. FIG. 23. FIG. 24.

FIGS. 20–24.—20. Hind-limb of man. 21. Monkey. 22. Dog. 23. Sheep. 24. Horse.

the knee is far removed and half-way down the limb (Figs. 20, 21). This is undoubtedly the original and normal condition of land-animals. But in all the more highly specialized and swifter animals the knee is brought nearer and nearer to the body, until, in the swiftest of all, such as the ruminants and the horse (Figs. 23, 24), it is high up on the side of the body, in the middle of what is usually called the thigh but which really includes the thigh and the upper part of the lower leg or shank.

2. See, again, the position of the heel. In man, monkey, bear, and many other mammals, and all *living* reptiles, the heel is on the ground, the tread is on the whole foot, plantigrade ; while in all the more special-ized and agile animals, and especially in the swiftest of all, such as the horse, the deer, etc., the heel is high in the air, and the tread is digitigrade.

3. Observe, again : there are two degrees of digiti-gradeness. The one we find in carnivorous or clawed digitigrades, the other in herbivores or hoofed digiti-grades. In the one the tread is on the whole length of the toes to the balls, as in man when he *tip-toes ;* in the other the *tread is on the tip of the last joint alone.* All that in any animal corresponds to the foot of a man—i. e., from the hamstring and heel downward—is called, in comparative anatomy, the "*pes.*" The pes, or foot of a horse, is eighteen inches long. It is easy to see what spring and activity this mode of treading gives to an animal. Think how helpless a horse would be if he trod on the whole foot, heel down !

10

4. Observe, again, the number of toes. In the process of specialization there is a tendency for these to become fewer and stronger.* The normal number, as already seen, is five. All the earliest mammals, and many orders of mammals still living, have five ; but in the most specialized orders, such as the ungulates or hoofed animals, they were steadily reduced in number in the course of evolution. In the elephant there are still five, in the hippopotamus there are four, in the rhinoceros three, in the goat two, in the horse one. Still more the order of the dropping is regular. If an animal have but four toes, it is usually the first, or great toe, or thumb, that is wanting, or may be rudimentary.

---

* This is only one example under a general law which it may be well to stop a moment to illustrate. A repetition of similar parts performing the same function is always an evidence of low organization, and as we rise in the scale of organization such parts usually become fewer and more efficient. Thus, to give one example, myriapods, as their name indicates, have hundreds of locomotive organs—lower crustaceans perhaps thirty or forty. As we go up, they are reduced to fourteen (tetradecapods), then to ten (decapods), then in spiders to eight, in insects to six, in vertebrates to four, and in man to two. A similar reduction in number, but increase in efficiency, is found in toes, *when they are used for support and locomotion only.* In man we find the normal number of five (1), because his hands are used for grasping and the functions of the fingers are not the same ; and (2), because man's development was almost wholly *brainward.* In other respects his structure is far less specialized than most other mammals. He can not compete with carnivores in strength and ferocity, nor with herbivores in fleetness. In the struggle for life, therefore, there was nothing left for him but increase in intelligence. Probably four is the smallest number of locomotive organs consistent with highest efficiency. In retaining but two legs for locomotion, man has lost in locomotive efficiency, but by the sacrifice he liberates two limbs for higher functions.

If, as in the rhinoceros, there are only three, then No. 5, or little toe, is also wanting, and the existing toes are Nos. 2, 3, and 4. If an animal has only two toes, as the goat, these are Nos. 3 and 4; and if only one, as the horse, it is the third or middle toe. Or, to put it more definitely : hoofed animals are divided into two groups, even-toed (artiodactyl) and odd-toed (perissodactyl). The even-toed may have four, as in the hippopotamus ; or two, as in the goat. The odd-toed may have three, as in the rhinoceros; or but one, as in the horse. Now, both of these orders came by differentiation, far back in the Eocene Tertiary, from a five-toed plantigrade ancestor. After dropping No. 1 (thumb or great toe) it is not yet decided, so far as number of toes is concerned, whether the resulting four-toed animal shall become artiodactyl or perissodactyl. If the former, then the two side-toes (Nos. 2 and 5) become shortened up, as in the hog ; then rudimentary, as in the ox and deer ; and finally pass away entirely, as in the goat. If, on the other hand, the four-toed animal is on the line of perissodactyl evolution, it becomes first a three-toed animal by dropping No. 5. Now, the two side-toes (Nos. 2 and 4) shorten up more and more, and the middle toe increases in size, until finally, in the modern horse, only the greatly enlarged middle toe (No. 3) remains. We look with wonder and admiration at the *danseuse* pirouetting on the point of one toe. The horse is performing this feat all the time. Yes, the one toe of a horse has all the three joints like ours.

The coffin-bone is the last joint, and the hoof is the nail.

Genesis of the Horse.—Every step of this process on the perissodactyl line may be traced in the history of the genesis of the horse. The beautiful form and structure of this animal were not made at once, but by a slow process of integration of small changes from generation to generation, and from epoch to epoch of the earth's history. The horse (as in fact did all ungulates) came from a five-toed *plantigrade* ancestor, but we are not able to trace the direct line of genesis quite so far. The earliest stage that we can trace with certainty, in this line of descent, is found in the eohippus of Marsh. This was a small animal, no bigger than a fox, with three toes behind and four serviceable toes in front, with an additional fifth palm-bone (splint), and perhaps a rudimentary fifth toe like a dew-claw. This was in early Eocene times. Then, in later Eocene, came the orohippus, which differs from the last chiefly in the disappearance of the rudimentary fifth toe and splint. (See Fig. 25.) Next, in the Miocene, came the mesohippus and miohippus. These were larger animals (about the size of a sheep), and had three serviceable toes all around; but in the former the rudiment of a fourth splint in the fore-limb yet remained. Then, in the Miocene, came the protohippus and pliohippus. These were still larger animals, being about the size of an ass. In the former the two side-toes were shortening up and the middle toe becoming larger. In the latter the two side-toes have become

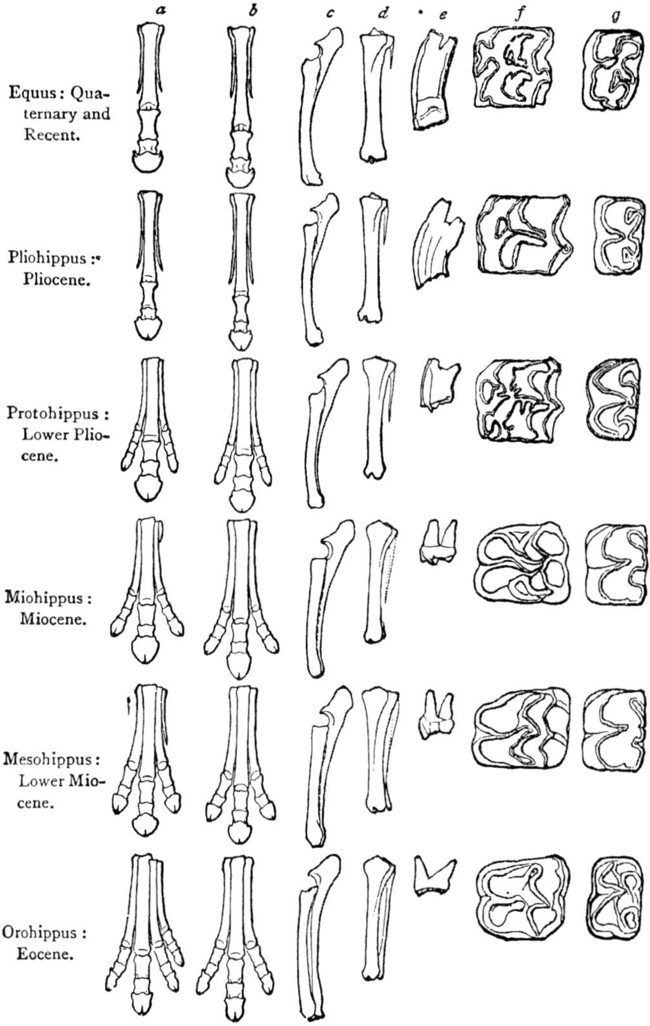

FIG. 25.—Diagram illustrating gradual changes in the horse family. Throughout *a* is fore-foot; *b*, hind-foot; *c*, fore-arm; *d*, shank; *e*, molar on side-view; *f* and *g*, grinding surface of upper and lower molars (after Marsh).

splints. Lastly, only in the Quaternary comes the genus *Equus*, or true horse. The size of the animal is become greater, the middle toe stronger, the side-splints smaller; but in the side-splints of the modern horse we have still remaining the evidence of its three-toed ancestor.

Similar gradual changes may be traced in the two leg-bones, which have gradually consolidated into one; in the teeth, which have become progressively longer and more complex in structure, and therefore a better grinder; in the position of the heel and wrist, which have become higher above-ground; in the general form, which has become more graceful and agile; and, lastly, in the brain, which has become progressively larger and more complex in its convolutions— to give greater battery-power, to make a more powerful dynamo—to work the improved skeletal machine. See, then, how long it has taken Nature to produce that beautiful finished article we call the horse !

We have taken only limbs as examples of what is true of the whole skeleton. To the superficial observer the bodies of animals of different classes seem to differ fundamentally in plan—to be entirely different machines, made each for its own purposes, at once, out of hand. Extensive comparison, on the contrary, shows them to be the same, although the essential identity is obscured by adaptive modifications. The simplest, in fact the only scientific, explanation of the phenomena of vertebrate structure is the idea of a primal vertebrate, modi-

fied more and more through successive generations by
the necessities of different modes of life.

See, then, in conclusion, the difference between man's
mode of working and Nature's.   A man having made a
steam-engine, and desiring to use it for a different purpose from that for which it was first designed and used,
will nearly always be compelled to add new parts not
contemplated in the original machine.   Nature rarely
makes new parts—never, if she can avoid it—but, on the
contrary, adapts an old part to the new function.   It is
as if Nature were not free to use any and every device to
accomplish her end, but were conditioned by her own
plans of structure ; as, indeed, she must be according to
the derivation theory.   For example : In early Devonian
times fishes were the only representatives of the vertebrate type of structure.   The vertebrate machine was
then a *swimming-machine*.   In the course of time, when
all was ready and conditions were favorable, reptiles were
introduced.   Here, then, is a new function—that of locomotion on land.   We want a *walking-machine*.   Shall
we have a new organ for this new function ?   No : the
old swimming-organ is modified so as to adapt it for
walking.   Time went on, until the middle Jurassic, and
birds were introduced.   Here is a new and wonderful
function, that of flying in the air.   We want a *flying-
machine*.   We know how man would have done this ; for
we have the result of his imagination in angels of Christian art and griffins of Greek mythology.   He would
have added wings to already existing parts, and this

would have necessitated the alteration of the whole plan of structure, both skeletal and muscular. Nature only modifies the fore-limbs for this new purpose. If we must have wings, we must sacrifice fore-legs. We can not have both without violating the laws of morphology. Finally, ages again passed, and, when time was fully ripe, man was introduced. Now we want some part to perform a new and still more wonderful function. We want a *hand*, the willing and efficient servant of a rational mind. We know, again, how man would have done this, for we have the result in the centaurs of Greek mythology, in which man's chest, and arms, and head are added to the body of a quadruped. But natural laws must not be violated, even for man. If we want hands, we must sacrifice feet. Again, therefore, the fore-limbs are modified for this new and exquisite function. Thus, in the fin of a fish, the fore-paw of a reptile or a mammal, the wing of a bird, and the arm and hand of a man, we have the same part, variously modified for many purposes.

Many other illustrations might be taken from the skeleton and from other systems, especially the muscular and nervous. But in the muscular system the modifications have been so extreme that homology is much more difficult to trace, and therefore requires more extensive knowledge than we yet possess, and more extended comparison than has yet been attempted. It has been traced with some success through mammals, and probably will be through air-breathing vertebrates—i. e., also through birds, reptiles, and amphibians ; but to trace it through fishes seems almost hopeless. In the case of the nervous system, and especially of the brain, it is again distinct; but this had better be taken up under another head, viz., proofs from ontogeny, Chapter VI.

In the visceral organs homology is very plain, in fact too plain. There is not modification enough in most cases even to obscure it, because function is the same in all animals. These organs do not, therefore, furnish good illustrations of that essential identity in the midst of adaptive modification which constitutes the argument for the derivative origin of structure. It is the organs of *animal life* that show this most perfectly, because it is these that take hold on the environment and are modified by it. There are, however, a few striking illustrations to be found among the visceral organs, especially the blood-system. This, however, had better also be deferred to the chapter on ontogeny.

CHAPTER VI.

WE have taken the vertebrate skeleton first, only because this department is most familiar. But in reality, the most beautiful illustrations of essential identity of structure in the midst of infinite diversity of adaptive modification for different functions and habits of life, and therefore of common origin from a primal form, are found in the department of articulates. I use the old Cuvierian department *articulata*, rather than the more modern *arthropods*, because the former includes worms also. Now, whether worms should be thus included with arthropods, or deserve a whole department to themselves it matters not for our purposes. It is generally admitted that arthropods probably descended from marine worms. They all have the same general plan of skeletal structure. It will suit my purpose, therefore, to regard worms as the lowest form of jointed animals.

Here, then, we have an entirely different plan of structure—a different style of architecture and different mechanical principles of machinery. Instead of a skeleton within and muscles acting on the outside, we have the skeleton on the outside, and muscles acting from within. Instead of two cavities, a neural and visceral, the skeleton forms but one cavity, in which all organs are inclosed and protected. Instead of finding the nerve-axis on the dorsal aspect of the body, we find it on the ventral aspect.

Take any articulate animal, for example, a shrimp, a centiped, or a beetle. Cut it across the body, and look at the end (Fig. 26). We see a ring of bone (chitin) in-

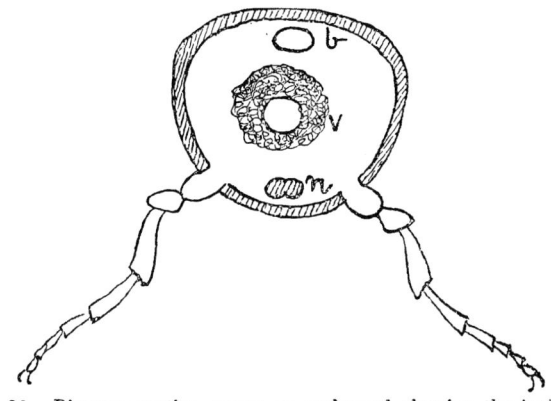

Fig. 26.—Diagram section across an arthropod, showing the inclosing skeleton-ring and a pair of jointed appendages. *n*, nervous center; *v*, viscera; *b*, blood system.

closing all the organs (nervous system *n*, blood system *b*, and visceral system *v*), and a pair of jointed appendages, perhaps legs, on each side. Now imagine these parts repeated in a linear series. The rings repeated make a hollow, jointed tube or barrel, the appendages repeated make a continuous row of appendages on each side. Now this is exactly what we actually find. The whole articulate skeleton is ideally made up of a series of such repeated rings and appendages, modified according to the position in the series, and the uses to which they are put. And then the whole articulate department is made up of such articulate animals again modified according to place in the scale of articulates. The modification in the lower forms is slight, and therefore the identity of the repeated parts is obvious ; but as we go up the scale, and the number and complexity of the functions increase, the adaptive modification becomes greater and greater, until finally it so obscures the essential identity, that it requires the most extensive comparison in the taxonomic series and in the ontogenic series, to pick up the intermediate links and establish the fact of common origin. In a word, whether they so originated or not, it is certain that the structure of articulate animals is exactly such as would be the case if all these animals were genetically connected, and came originally from a primal form something like one of the lower crustaceans, or, perhaps, a marine worm.

It will be best to take an example from about the middle of the scale, where the two elements, viz., essential identity and adaptive modification, are somewhat evenly balanced, and both traceable with ease and certainty. Take, then, a cray-fish, a lobster, or a shrimp. This animal (Fig. 27) has twenty or twenty-one rings and pairs of jointed appendages. The rings are some of them diminished, some of them increased in size.

Sometimes several are consolidated; sometimes several are partially or wholly aborted. The appendages are modified in shape and size, according to their position, so as to make them swimming-appendages (swimmerets), walking - appendages (legs), eating-appendages (jaws), and sense-appendages (antennæ).

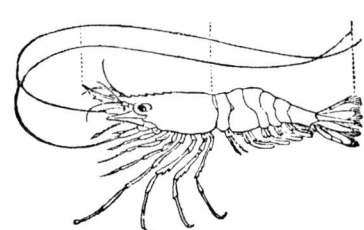

Fig. 27.—Shrimp (Palæmonetes vulgaris).

For example, in the abdominal region, or so-called tail, we have seven segments, all being perfect movable rings, each with its pair of jointed appendages, except the last, or *telson*. The appendages of the first ring (Fig. 28, B) are specially modified in the male as organs of copulation (B'). The next four pairs are modified for swimmerets (D') and for use as holders of the eggs in the female. The appendages of the sixth ring (G) are broad and paddle-shaped, and, together with the telson or seventh ring (H), form the powerful terminal swimmer. Going, now, to the cephalo-thorax : in this either a large number of segments (thirteen or fourteen) are consolidated above to form the upper shell or carapace ; or else, as is more probable, two or three of the anterior segments have enlarged and grown backward over, and at the expense of the others, to form this shell. At any rate, it is certain that the carapace is formed of the dorsal portions of a number of segments consolidated

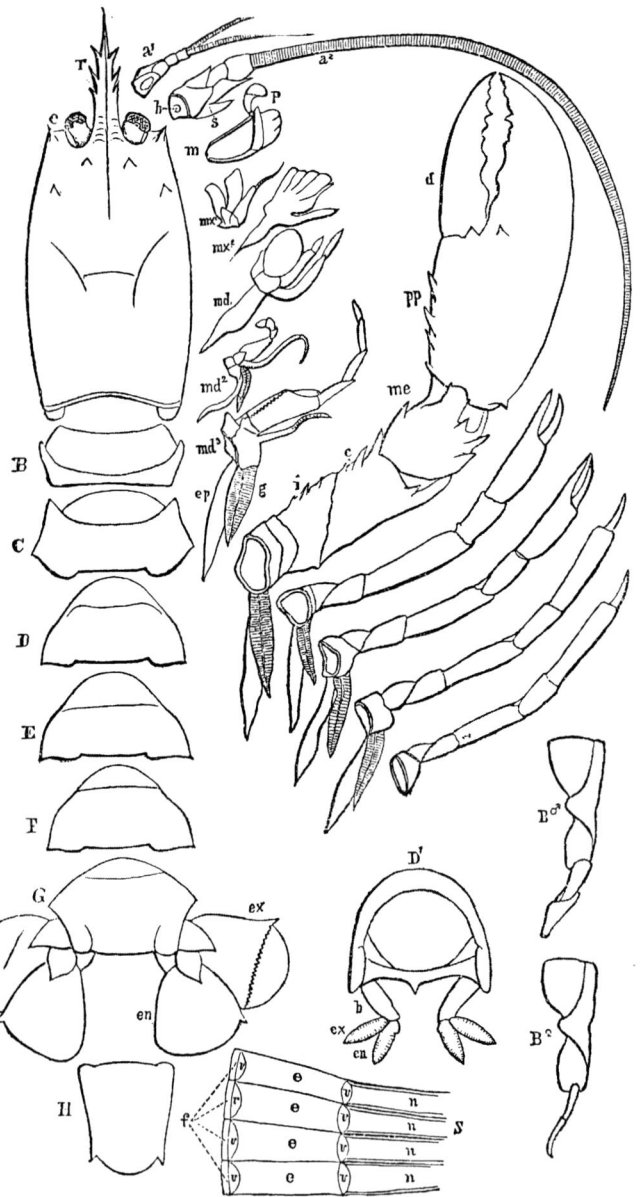

Fig. 28.—External anatomy of the lobster (after Kingsley).

together. Below, however, the segments are all distinct, and have each its own pair of appendages. For example, going forward in this region, the five next pairs of appendages are greatly enlarged and very strong, and serve the purpose of locomotion. They are *walking - appendages*. The next two or three pairs are smaller and somewhat modified, but not so much as to obscure their essential similarity to legs. Like legs, they are many-jointed, and like legs, too, they

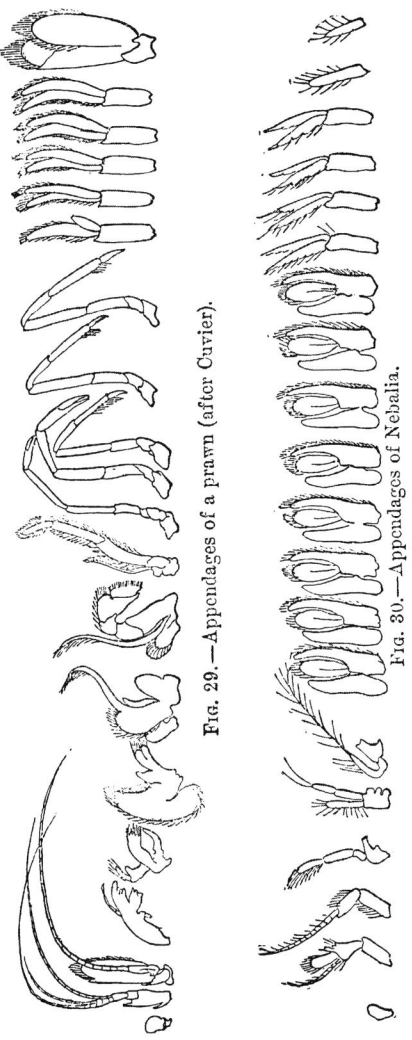

Fig. 29.—Appendages of a prawn (after Cuvier).

Fig. 30.—Appendages of Nebalia.

have gills attached to them. They are called maxilli-peds, or jaw-feet. They are used like hands to gather food and carry it to the mouth. They are *gathering-appendages*. Then follow three or four pairs still more modified, and used for mastication. They are called maxillæ and mandibles. They are *eating-appendages*. Then follow two pairs, long, many-jointed, with the same kind of curious hinge-joints, which we have in the legs, undoubtedly homologous with all the others, but used for an entirely different purpose, and special-ly modified for that purpose. They are the antennæ. They are delicate organs of touch and of hearing, for

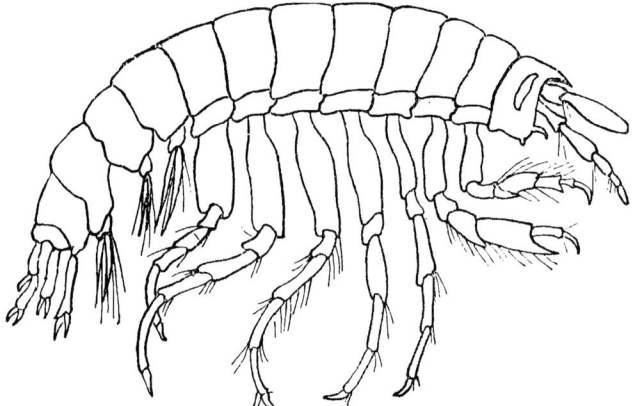

Fig. 31.—Vibilia, an amphibod crustacean (after Milne Edwards).

the ear is situated in the basal joint of the anterior pair. Last of all, there is still another pair, jointed and movable, on the ends of which are situated the eyes. These last three, therefore, are *sense-appendages*. Some writers make this last pair special organs, not homologous with appendages.

For the sake of greater distinctness, we give the whole series of these appendages in one of the higher forms, viz., the prawn (Palemon, Fig. 29, and in one of the lower forms, Nebalia, Fig. 30).

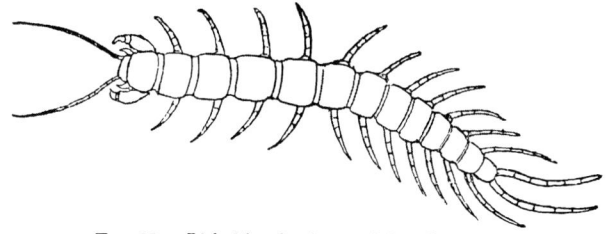

FIG. 32.—Lithobius forcipatus (after Carpenter).

That these are really homologous parts is further shown by the fact that in the case of other crustaceans, such as limulus, the same appendages, i. e., the appendages of the same body segments, which in the cases before mentioned are used as feet, become swimmers, while the appendages corresponding to jaw-feet become walkers ; and even what corresponds to antennæ or sense-appendages, may, as in branchippus, become powerful claspers. Finally, in all the lowest crustaceans, the identity is evident, because all the segments and their appendages are much alike in form and function (Fig. 31).

We have taken examples from near the middle of the articulate scale, because, as already stated, both the essential identity and the adaptive modifications are easily

11

traced. If we go downward in the scale, the structure becomes more and more generalized, and the rings and appendages become more and more alike (Fig. 31), until in the most generalized forms we have only a series of similar rings, with similar pairs of appendages, except some necessary modifications to form the head and tail. This is well shown in the centiped (Fig. 32), and still better in marine worms (Fig. 33). In some marine worms the slight modification to form the head takes

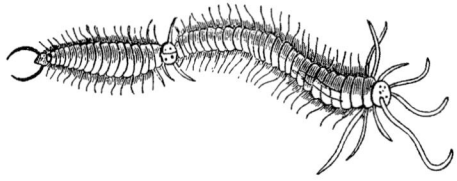

FIG. 33.—Syllis prolifera.

place under our very eyes. These often multiply by dividing themselves into two. When they do so, they make a new head and new tail by slight modification of segments and appendages (Fig. 33).

If, on the other hand, we go up the scale, we find adaptive modifications obscuring more and more the simple and obvious identity of parts, until finally the identity can not be recognized without extensive comparison in the taxonomic series and study of embryonic conditions. In crabs—which is a higher form than crayfish—the tail or abdomen seems to be wanting, but is only very small and bent under the body and thus concealed. In all essential respects the structure is precisely like the cray-fish. In fact, in the embryo, we trace the one form into the other; for the crab is at first a long-tailed crustacean (Fig. 34).

Insects are the highest form of articulates. In these, therefore, we find the modification is still greater than in

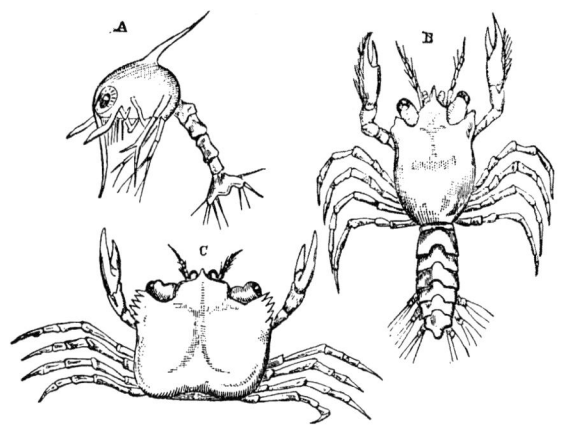

FIG. 34.—Development of Carcinus mœnas. A, zoæa stage; B, megalopa stage; C, final state (after Couch).

crustaceans, though even here the ring-and-appendage structure is plain enough in most cases.

One of the best evidences of high grade among animals is the gathering of the segments into distinct groups, and especially the distinctness of the *head* as one of these groups. In worms and lower crustaceans there is no grouping at all, the skeleton being a continuous series of joints, only slightly modified at the anterior and posterior extremities. In the higher crustacea, and in spiders and scorpions, they are grouped into two

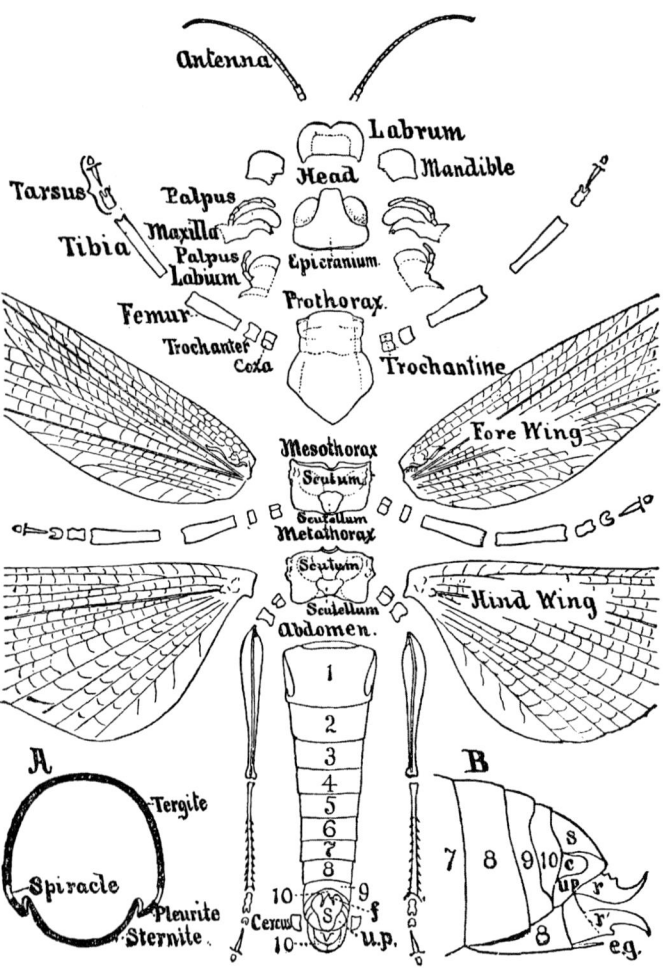

FIG. 35.—External anatomy of Caloptenus spretus, the head and thorax disjointed; up, uropatagium; *f*, furcula; *c*, cercus (drawn by J. T. Kingsley).

regions, viz., cephalo-thorax and abdomen. In insects
they are grouped into three very distinct regions—head,
thorax, and abdomen. In insects, therefore, we find for
the first time the head distinctly separated from the rest
of the body. This is an evidence of high grade, because
it shows the dominance of head-functions.

The insect, such, for example, as a grasshopper, consists of seventeen or eighteen segments (Fig. 35). Of
these, four belong to the head, three to the thorax, and
about ten to the abdomen. Those of the abdomen are
all separated and movable ; those of the thorax and head
are more or less consolidated. The appendages of the
head-segments become antennæ and jaw-parts, i. e., mandibles—maxillæ and labium ; the appendages of the thorac-
segments become legs (the wings are not homologous
with appendages), while those of the abdomen are aborted.
The steps of the gradual consolidation on the one hand,
and the abortion on the other, may be traced in the embryo or larva—i. e., in the caterpillar or the grub of a
bee or a beetle. In the caterpillar, for example, there is
no grouping into three regions, there is no consolidation, and all the segments have appendages. Again, the
almost infinite variety in the mouth-parts among insects, brought about by adaptive modifications for biting,
for piercing, and for sucking, and yet the essential identity of all to the more simple and generalized structure
of the grasshopper, is an admirable illustration of the
same principle. But to dwell upon these minor points
would carry us too far.

Illustration of the Law of Differentiation.—We have here, in the modifications of segments and appendages of articulates, an admirable illustration of the most fundamental law of evolution, viz., the law of differentiation. As we have already seen (page 21), perhaps the most beautiful and certainly the most fundamental illustration of this law is found in the development of cell-structure. Commencing in the lowest animals, and in the earliest embryonic stages of the higher animals, from a condition in which all are alike, the *cells* as we go upward quickly diverge into different forms to produce different tissues and perform different functions. Here, then, we have a perfect example of essential identity and adaptive modification. It is the very best type of differentiation. So also skeletal *segments*, commencing, in the lowest articulates and in earliest embryonic stages of the higher, all alike, as we go upward in either series, begin immediately to diverge in various directions (divergent variation), taking different forms to subserve different uses. Here, again, therefore, is an illustration of the law of differentiation. Lastly, in the articulate department, commencing with the lowest forms and earliest embryonic conditions, and we may add earliest geological times, and going up either series from generalized forms very much alike, the *individuals* are gradually differentiated into many special forms, in order to adapt them to the diversified modes of life actually found in nature. Thus cells, segments, individuals, are all alike affected by this most fundamental law.

We have taken our illustrations from only the two departments of vertebrata and articulata, because these are the most familiar to the reader, and also have been most carefully studied. We have shown that the general structure of all vertebrates is precisely what it would be if they all had come from one primal vertebrate form, and that of all articulates what it would be if all had come from one primal articulate form. The only *natural* explanation, and, therefore, the only scientific explanation of this, is that *they were really thus derived*. The same kind of evidence may be drawn from the study of other departments, but to pursue the subject any further in this direction would carry us beyond the limits which we have assigned. We desire only to explain the nature, not to give all, of the evidence. The examples given will be sufficient for the purposes of illustration. The whole proof is nothing less than the whole science of comparative anatomy.

Vertebrates, then, were derived from a primal vertebrate, articulates from a primal articulate, and so for other departments. But whence were these *primals* derived? Are there any intermediate links between, any deeply concealed common plan of structure underlying these primary groups, showing a common origin? It must be confessed that, in their *mature* condition, there seems to be but little evidence of such. These primary groups seem to be built on different plans, to be fundamentally of different styles of architecture. Therefore Darwin, in the true spirit of inductive caution—that true scientific spirit which keeps strictly within the limits of evidence —commences with four or five distinct primal kinds, from which by divergent variation all animals were descended. Nevertheless, the truly scientific biologist must ever strongly incline to believe that these also came from some *primal animal*, and even that both animals and plants were derived from some primal form of *living thing;* that as, in the taxonomic series, the animal and vegetal kingdoms in their lowest forms merge undistinguishably into one another ; as in the ontogenic series the animal and plant germ are one, so also in the phylogenic series the earliest organisms were simply living things, but not distinctively animal nor vegetal. Science, therefore, whose mission is to trace origins as far back as possible, must ever strive to find connecting links between the primary groups. Some such have been supposed to have been discovered. Some find the origin of vertebrates among the molluscoids (ascidians) ; some find the origins of both vertebrates and articulates among marine worms (annelids). This point is still too doubtful to be dwelt upon here. It may be that we seek in vain for such connecting links among existing forms. It may well be that the point of separation of these great primary groups (unless we except vertebrates) was far lower even than these low forms. Both phylogeny and ontogeny seem to indicate this. In the earliest fauna known, the primordial (for if there was life in the archæan it was not yet differentiated into a fauna), all the great departments, except the vertebrates, seem to have been represented. In embryonic development, too, the point of connection or even of similarity, between the great departments, is found, as we shall see hereafter, only in the earliest stages—i. e., lower down than any but the lowest existing forms, viz., the protozoa.

CHAPTER VII.

IT is a curious and most significant fact that the successive stages of the development of the *individual* in the higher forms of any group (ontogenic series) resemble the stages of increasing complexity of differentiated structure in ascending the animal scale in that group (taxonomic series), and especially the forms and structure of animals of that group in successive geological epochs (phylogenic series). In other words, the individual higher animal in embryonic development passes through temporary stages, which are similar in many respects to permanent or mature conditions in some of the lower forms in the same group. To give one example for the sake of clearness : The frog, in its early stages of embryonic development, is essentially a fish, and if it stopped at this stage would be so called and classed. But it does not stop ; for this is a temporary stage, not a permanent condition. It passes through the fish stage and through several other temporary stages, which we shall explain hereafter, and onward to the highest condition attained by amphibians. Now, if we could trace perfectly the successive forms of amphibians, back through the geological epochs to their origin in the Carboniferous, the resemblance of this series to the stages of the development of a frog would doubtless be still closer. Surely this fact, if it be a fact, is wholly inexplicable except by the theory of derivation or evolution. The embryo of a higher animal of any group passes *now* through stages represented by lower forms, because in its evolution (phylogeny) its ancestors *did actually have these forms*. From this point of view the ontogenic series (individual history) is a brief recapitulation, as it were, from memory, of the main points of the philogenic series, or family history. We say brief recapitulation of the *main* points, because many minor points are dropped out. Even some main points of the earliest stages of the family history may be dropped out of this sort of inherited memory.

This resemblance between the three series must not, however, be exaggerated. Not only are many steps of phylogeny, especially in its early stages, dropped out in the ontogeny, but, of course, many adaptive modifications for the peculiar conditions of embryonic life are added. But it is remarkable how even these—for example the umbilical cord and placenta of the mammalian embryo—are often only modifications of egg-organs of lower animals, and not wholly new additions. It is the similarity in spite of adaptive modifications that shows the family history.

We will now illustrate by a few striking examples.

We can not do better than to take, again, as our first example, the development of tailless amphibians, and dwell a little more upon it :

1. **Ontogeny of Tailless Amphibians.**—It is well known that the embryo or larva of a frog or toad, when first hatched, is a legless, tail-swimming, water-breathing, gill-breathing animal. It is essentially a fish, and would be so classed if it remained in this condition. The fish retains permanently this form, but the frog passes on. Next, it forms first one pair and then another pair of legs ; and meanwhile it begins to breathe also by lungs. At this stage it breathes equally by lungs and by gills, i. e., both air and water. Now, the lower forms of amphibians, such as siredon, menobranchus, siren, etc., retain permanently this form, and are therefore called *perennibranchs*, but the frog still passes on. Then the gills gradually dry up as the lungs develop, and they now breathe wholly by lungs, but still retain the tail. Now this is the permanent, mature condition of many amphibians, such as the triton, the salamander, etc., which are therefore called *caducibranchs*, but the frog still passes on. Finally, it loses the tail, or rather its tail is absorbed and its material used in further development, and it becomes a perfect frog, the highest order (*anoura*) of this class.

Thus, then, in ontogeny the fish goes no further than the fish stages. The perennibranch passes through the fish stage to the perennibranch amphibian. The caducibranch takes first the fish-form, then the perennibranch-form, and finally the caducibranch-form, but goes no

further.   Last, the anoura takes first the fish-form, then
that of the perennibranch, then that of the caducibranch, and finally becomes anoura.   This is shown in
the diagram, which must be read upward, line by line.

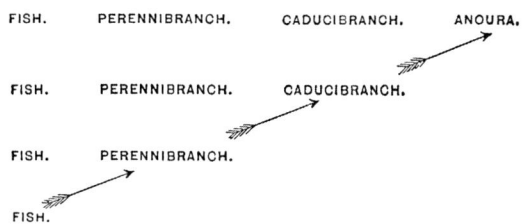

Diagram showing the stages of development of amphibians.   (To be read
upward.)

Now, this is undoubtedly the order of succession of
forms in geological times—i. e., in the phylogenic series.
This series is indicated by the arrows in the diagram.
Fishes first appeared in the Devonian and Upper Silurian
in very reptilian or rather amphibian forms.   Then in
the Carboniferous, fishes still continuing, there appeared
the lowest—i. e., most fish-like—forms of amphibians.
These were undoubtedly perennibranchs.   In the Permian and Triassic higher forms appeared, which were certainly caducibranch.   Finally, only in the Tertiary, so far
as we yet know, do the highest form (anoura) appear.
The general similarity of the three series is complete.
If we read the diagram horizontally, we have the ontogenic series ; if diagonally with the arrows, we have both
the taxonomic and the phylogenic series.

2. Aortic Arches.—But some will, perhaps, say that
these stages in the ontogeny are only examples of adapt-

ive modifications—like modifications for like conditions of life—and had better be accounted for in this way, without reference to family history. We will, therefore, take another example, which can not be thus accounted for—an example in which there is no possible use *now* for the peculiar form or structure which we find. For

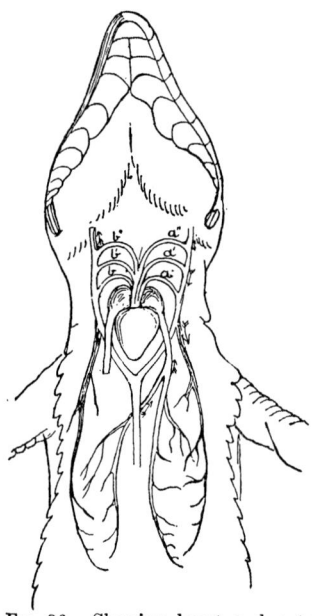

this purpose we take the case of the *course of circulation in vertebrates.*

If one examines the large vessels *going out* from the heart of a lizard, he will find *six aortic arches* — i. e., three on each side. These all unite below to form the one descending abdominal aorta. This is shown in the accompanying figure (Fig. 36), in which *a a' a''* and *b b' b''* are the six arches. Now, there is no conceivable use in having so many aortic arches. We know this, because there is but one in birds and mammals, and the circulation is as

FIG. 36.—Showing heart and outgoing blood-vessels of a lizard (after Owen). The arrows show the course of the blood.

effective, nay, much more effective in these than in reptiles. The explanation of this anomaly is revealed at once as soon as we examine the circulation of a *fish*, which is shown in the accompanying figure (Fig. 37). The multiplication of the aortic arches is here, of

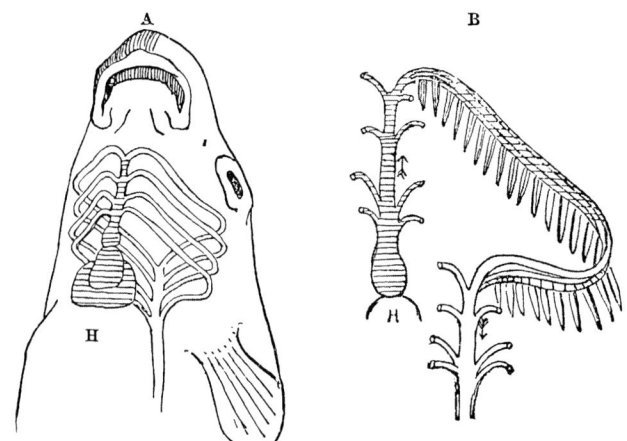

FIG. 37.—A, heart and gill-arches of a fish; B, one arch with fringe (after Owen); H, the heart.

course, necessary, for they are the *gill-arches*. The whole of the blood passes through these arches, to be aërated in the gill-fringes. The use of this peculiar structure is here obvious enough. If a lizard were ever a fish, and afterward turned into a lizard, changing its gill-respiration for lung-respiration, then, of course, the useless gill-arches would remain to tell the story. Now, although a lizard never was a fish, in its *individual* history or ontogeny, it was a fish in its family history or phylogeny, and therefore it yet retains, by heredity, this curious and *useless* structure as evidence of its ancestry.

That this is the true explanation is demonstrated by the fact that in amphibians this very change actually takes place before our eyes in the *individual history*. We have already seen that the individual frog, in its tadpole state, is a gill-breather. It has therefore its gill-arches

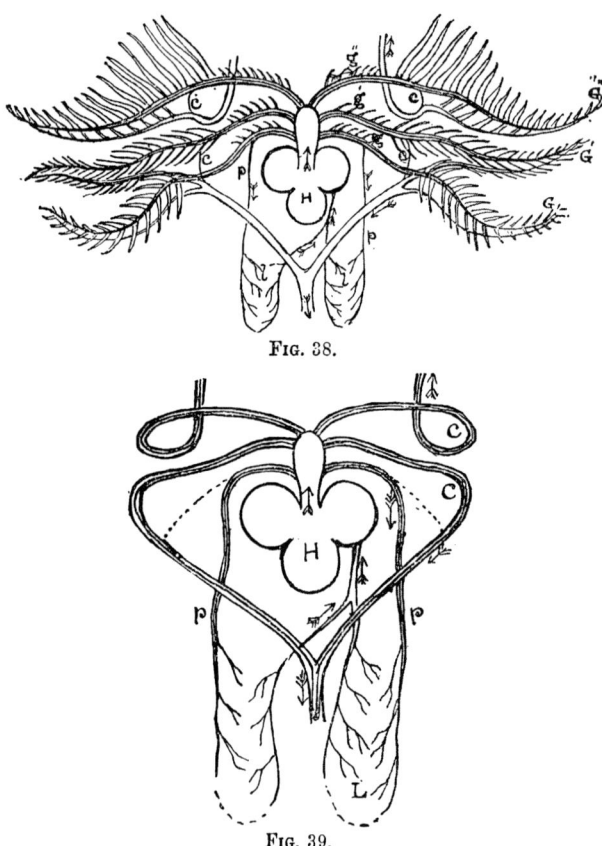

FIG. 38.

FIG. 39.

FIGS. 38, 39.—Diagrams showing the change of the course of blood in the development of a frog. 38. The tadpole stage. 39. The mature condition. H, heart; G G'G'', external gills; *g g'g''*, internal gills; *c c*, connecting branches in the tadpole; *p p*, pulmonary branches.

(Fig. 38), three on each side, like a fish, and for the same reason, viz., the aëration of the blood. But when its gills dry up and lung-respiration is established, its now useless gill-arches still remain as aortic arches, to attest their previous condition (Fig. 39). Now, the lizard undoubtedly came from an air-breathing, tailed amphibian, and therefore inherited this form of arterial distribution. In both lizard and amphibian the ultimate cause is an origin from fishes, in which such arches are obviously necessary. The diagrams, Figs. 38 and 39, are illustrations somewhat idealized, showing the manner in which the change actually takes place in air-breathing amphibians. Fig. 38 represents the tadpole stage, and Fig. 39 the mature condition. In the former the gills are mostly external, G G', etc., but also internal, $g g'$, as in the fish. Observe in this condition the small connecting vessels, $c c'$. When the external gills dry up, these are enlarged, and the whole of the blood passes through them, as shown in Fig. 39. It is seen, also, in Fig. 38, that a small branch, p, goes from the lower gill-arches to the yet rudimentary lung, l. When the gill-fringes have disappeared, the whole of the blood of the lower arch goes through the now enlarged pulmonary branch to the lungs, L, now in full activity, and the remainder of this arch disappears, as shown by the dotted lines in Fig. 39.

The change which actually took place in the family history of the lizard probably differed from the above only in being more simple, the gills being only internal like the fish. The external gills complicate the process

12

a little in the case of the frog, but the principle is precisely the same.

As already explained (pages 82–85), the large gap between fishes and reptiles, as regards mode of respiration, is completely filled both in the taxonomic series —i. e., in ganoids, dipnoi, and the mature condition of the different orders of amphibians—and in the ontogeny of the higher amphibians. Now, we add that the same is true of the arterial distribution. We have just traced the change in the ontogeny of the frog, but the steps of the same change are traceable in passing from the typical fish (teleosts), through dipnoi and amphibians to reptiles. Thus, again, the phylogeny, the taxonomy, and the ontogeny, are in complete accord.

But the argument for evolution does not stop here. If birds and mammals have come from reptiles, and therefore from fishes, we may expect to find some evidences of the same kind still lingering in the great arteries. And such we do find. It is a most curious and significant fact that, in the early embryonic condition of birds and mammals, including man himself, we find on each side of the neck several gill-slits, each with its gill-arch, and therefore *several aortic arches on each side*, precisely similar to what we have already described. These arches are subsequently, some of them, obliterated ; some modified to form the one aortic arch, and some of them still more modified to form the other great arteries coming from the heart to supply the head and forelimbs.

This is so beautiful and convincing an example, and one so generally unfamiliar, to even intelligent persons,

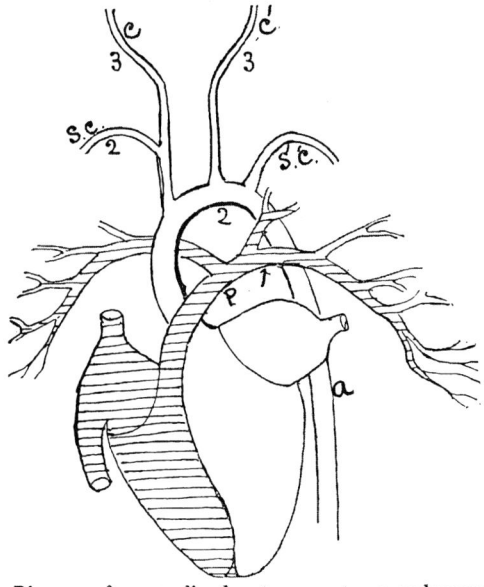

FIG. 40.—Diagram of mammalian heart. *a*, aorta ; *p*, pulmonary artery ; *scs'c'*, subclavium on each side ; *cc'*, carotids on each side.

not especially acquainted with biology, that it is best to explain it more fully. In Fig. 40 we give a mammalian heart and outgoing vessels, very slightly modified, so as to suggest the process of change. In Fig. 41 we give an ideal diagram representing the primitive aortic arches as they exist in the embryo of mammals, birds, and reptiles. It represents, also, substantially, the arches as they exist in the *mature* condition in the most reptilian fishes (dipnoi) and in some sharks, except that in these the arches are of course furnished with gill-fringes. We will use this figure, therefore, to represent both the em-

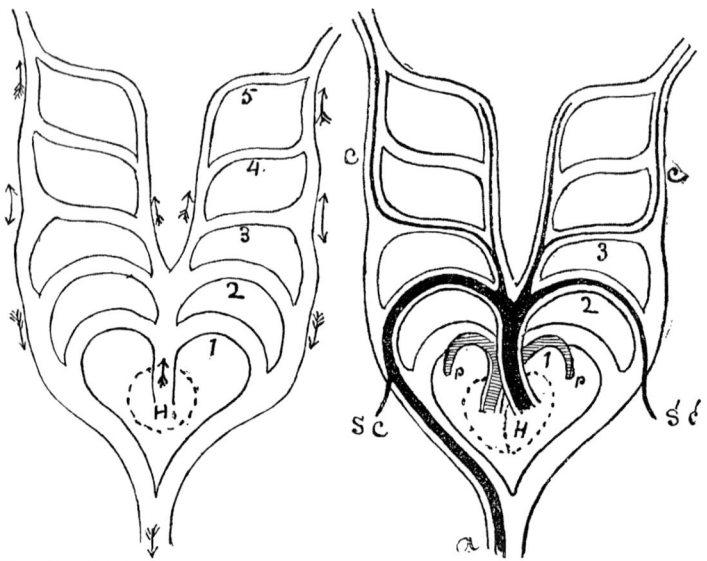

FIG. 41.—Ideal diagram representing the primitive aortic arches (after Rathke).

FIG. 42.—Modified for bird.

bryonic condition of air-breathing vertebrates and the mature condition of some fishes. The place of the heart is indicated by the dotted circle. Fig. 36, on page 134, shows what these arches become in reptiles (lizard). It is seen that the two upper arches on each side are obliterated, as indeed they already are in some teleost fishes. Fig. 42 shows what they become in birds. The two upper arches are, of course, obliterated. The others are all modified, each in a manner which may be readily understood by comparison with Fig. 41.   Finally, Fig. 43 shows what they become in mammals and in man.   In the bird (Fig. 42) the first pair of arches become the two pulmonary arteries as they do also in the lizard.   The second pair become on the right side (left of the diagram) the aortic arch, on the left side (right of the diagram) the left subclavian, $s'c'$ (the right subclavian, sc, is a branch of the aortic arch).   The third pair become carotids, cc, while the fourth and fifth, as already said, are aborted.   In the mammal (Fig. 43), on the left side (right of the diagram) the first arch becomes the pulmonary

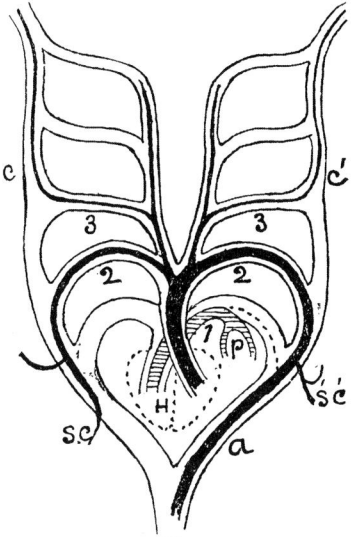

Fig. 43.—Modified for mammal.

artery, p.   In the fœtus the continuation of this arch forms the ductus arteriosus, which is afterward obliterated, as shown in the dotted line.   The second arch becomes the aortic arch, the third the left exterior carotid. On the right side (left of the diagram) the first arch becomes aborted ; the second, the right subclavian, sc (the *left* subclavian, $s'c'$, is a branch of the aortic arch) ; and the third, the right carotid.   Nos. 4 and 5, on both sides, as usual, are aborted.

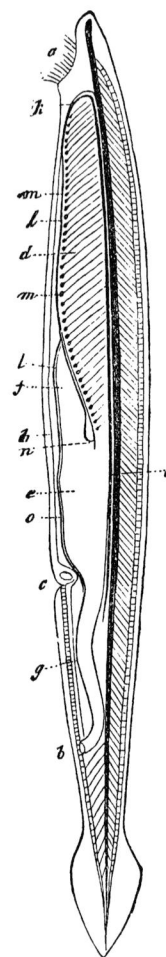

Fig. 44. — Lancelet (Amphioxus lanceolatus). Magnified two and one-half times.

See, then, the gradual process of change through the whole vertebrate department. In the lowest of all vertebrates, if vertebrate it may be called (for what corresponds to its backbone is an unjointed, fibrous cord), the amphioxus or lancelet (Fig. 44), there are about forty gill-arches on each side. As we rise in the scale of fishes these are reduced in number. In the lamprey, there are seven ; in the sharks, usually five ; in ordinary fishes (teleosts), there are four or sometimes only three on each side, the others being aborted. Thus far the change is only by diminution of number in accordance with a law universal in biology, that decrease in the number of identical organs is evidence of advance in the grade of organization, provided that it be associated with more perfect structure of the organ. The further change is one of adaptive modification. In some reptiles (lizard) the three gill-arches on each side all retain the form of aortic arches ; in some reptiles only two retain this form. In birds and mammals only one arch is retained, in the form of aortic arch, the others being modified to form the great outgoing vessels of the heart, or else aborted. It may be well to observe that in birds the one aortic arch turns to the right, while in mammals it turns to the left. This is positive evidence that mammals could not have come from birds, nor *vice versa*. They both came from reptiles, and, of the many reptilian arches, a right one was retained by the bird branch, and a left one by the mammalian.

In all the figures illustrating this subject, we have left out the great *incoming* vessels or veins, because we are not here concerned with them, they not being transformed gill-arches.

Last of all, it may be well to stop a moment to show the cogency of this evidence. If it were a question of the origin of some structure not only useful (for all structures selected by Nature must be useful) but the *best imaginable,* like the eye or the ear, for example ; then, if *we examined only the highest form* or *the finished article,* there are two ways in which it is possible to explain the adaptive structure. We may either suppose that it was made at once out of hand, by some intelligent contriver ; or else that it was slowly made by a process of evolution, becoming more and more perfect by a selection of only the most perfect from generation to generation. But in the case of the six aortic arches of the lizard, we are shut up to the one explanation only, viz., by slow process of evolution. One arch is all that is necessary, as is plainly shown by the use of only one in the more perfect circulation of birds and mammals. If the thing were done out of hand, unconditioned by the previous structure in fishes, to have made six was surely but a bungling piece of work.

3. **Vertebrate Brain.**—Another excellent example is the structure of the vertebrate brain. The brain of an average fish is represented in Fig. 45. It consists of four

FIG. 45.—Fish-brain. A, side view ; B, top view.

or five swellings, or ganglia, strung along, one beyond another. Commencing behind, these are, first, the medulla, *m ;* then the cerebellum, *cb ;* then the optic lobes, *ol ;* then the cerebrum and thalamus combined, *cr ;* and last, the olfactive lobes, *of.* Of these, it will be observed, the optic lobe is the largest in the brain of the fish (Fig. 45). In the brain of the reptile (Fig. 46) we have the

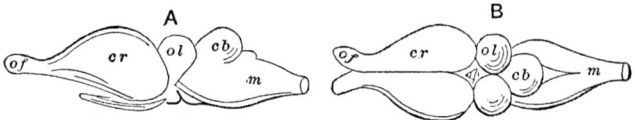

FIG. 46.—Reptile-brain. A, side view ; B, top view.

same serial arrangement, of the same parts, only that the cerebrum has now become the dominant part instead of the optic lobes. In the average bird (Fig. 47) the cerebrum has grown so large that it extends backward, and partly covers the optic lobes. In the lower mammals (marsupials), the brain is much the same in this respect, as in birds—i. e., the cerebrum only partly covers the optic lobes, so that, looked at from above, the whole se-

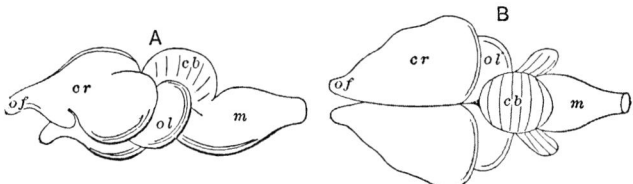

FIG. 47.—Bird-brain. A, side view ; B, top view.

ries of ganglia are still visible. But in the average mammal (Fig. 48) the cerebrum is so enlarged that it covers

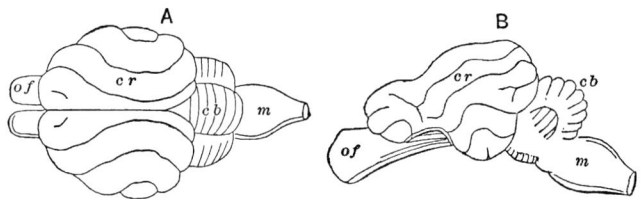

FIG. 48.—Mammal-brain. A, top view ; B, side view.

entirely the optic lobes and encroaches on the cerebellum behind and the olfactive lobes in front. In some monkeys, indeed, the cerebellum is nearly or even quite covered. Finally, in man (Fig. 49), the cerebrum has

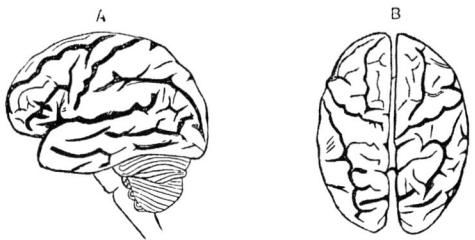

FIG. 49.—Man's brain. A, side view ; B, top view.

grown so enormously that it covers every other part and completely conceals them from view when the brain is looked at from above. In front it not only covers but

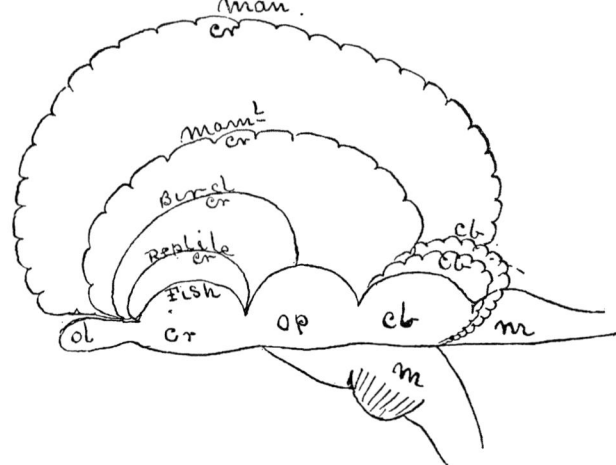

FIG. 50.—Ideal section showing all the above stages.

has grown far beyond the olfactive lobes ; behind it extends beyond and overhangs the cerebellum ; on the sides it overhangs and covers all. Looked at from above, nothing is seen but this great ganglion. The ideal section (Fig. 50) represents all these stages diagrammatically in one figure. After what has been said, the figure will be readily understood.

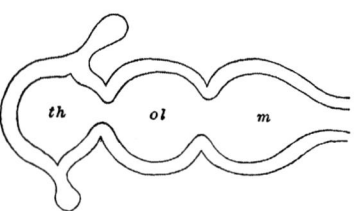

FIG. 51.—Sub-fish stage. *th*, thalamus ; *ol*, optic lobe ; *m*, medulla.

Now, it is a most remarkable fact that substantially these same stages, which are permanent conditions in the

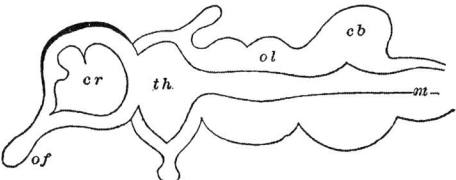

FIG. 52.—Fish-stage.   *of*, olfactive lobe; *cr*, cerebrum; *th*, thalamus; *ol*, optic lobe; *cb*, cerebellum; *m*, medulla.

taxonomic series, are passed through as transient stages in the embryonic development of the human brain, and in the order given above.   The very early condition of the human brain is represented in Fig. 51.   It is evidently

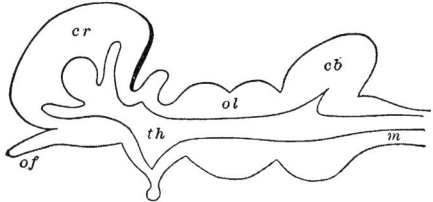

FIG. 53.—Reptile-stage.

nothing more than the intercranial continuation of the spinal cord, enlarged a little into three swellings or ganglia.   These are the early representatives of the medulla, the optic lobes, and the thalamus ; which last may be regarded as the basal and most fundamental part of the cerebrum.   This stage may be regarded as lower than that of the ordinary fish.   I have called it, therefore, the *sub-fish stage*.   The cerebellum is a subsequent outgrowth from the medulla, as is the cerebrum and olfactive lobes from the thalamus. Fig. 52 may be said, therefore, to represent fairly the fish-stage. Henceforward the principal growth is in the cerebrum and cerebellum, both of which are subsequent outgrowths of the original simple ganglia, the medulla, and the thalamus. The cerebrum especially increases steadily in relative size, first becoming larger than but not covering the optic lobes (Fig. 53). This represents the reptilian stage. Next, by

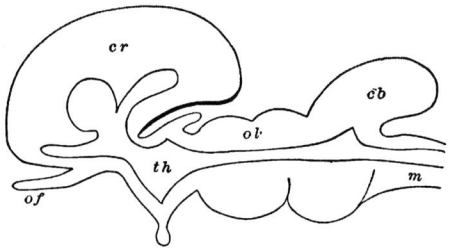

FIG. 54.—Bird-stage. *of*, olfactive lobe; *cr*, cerebrum; *th*, thalamus; *ol*, optic lobe; *cb*, cerebellum; *m*, medulla.

further growth, it covers partly the optic lobes (Fig. 54). This may be called the bird-stage. Then it

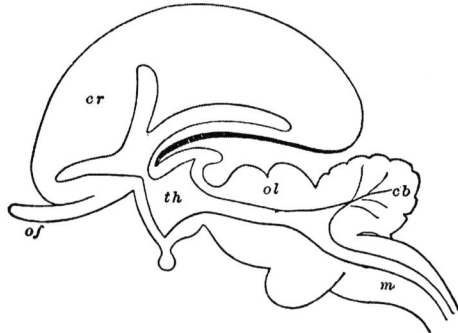

FIG. 55.—Mammalian stage.

covers wholly the optic lobes, and encroaches on the cerebellum behind and olfactive lobes in front (Fig. 55) This is the mammalian stage. Finally, it covers and

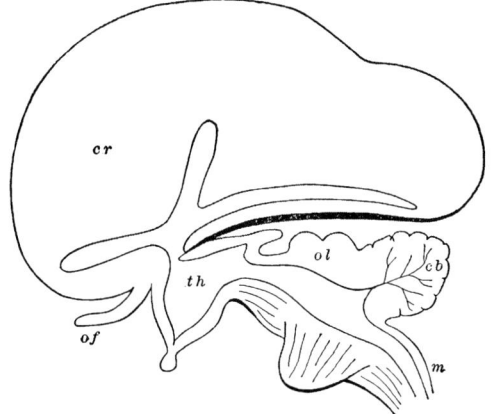

FIG. 56.—Human stage.

overhangs all, and thus assumes the human stage (Fig. 56).

We have spoken thus far only of relative *size;* but progressive changes take place also in complexity of structure—i. e., in the depth and number of convolutions of the cerebrum and cerebellum. The cerebrums of fish, of reptile, bird, and lower mammals are smooth. About the middle of the mammalian series it begins to be convoluted. These convolutions become deeper and more numerous as we go upward in the scale, until they reach the highest degree in the human brain. The object of these inequalities is to increase the surface of gray matter—i. e., the extent of the force-generating as compared with the force-transmitting part of the brain, or battery as compared with conducting-wire. Now, in embryonic development the human brain passes also through these stages of increasing complexity of organization. Here also the ontogenic is similar to the taxo-nomic series.

Now, why should this peculiar order be observed in the building of the individual brain ? We find the answer, the only conceivable scientific answer to this question, in the fact that *this is the order of the building of the vertebrate brain by evolution* throughout geological history. We have already seen that fishes were the only vertebrates living in the Devonian times. The first form of brain, therefore, was that characteristic of that class. Then reptiles were introduced ; then birds and marsupials ; then true mammals ; and, lastly, man. The different styles of brains characteristic of these classes were, therefore, successively made by evolution from earlier and simpler forms. In phylogeny this order was observed because these successive forms were necessary for perfect adaptation to the environment at each step. In taxonomy we find the same order, because, as already explained (page 11), every stage of advance in phylogeny is still represented in existing forms. In ontogeny we have still the same order, because ancestral characteristics are inherited, and family history recapitulated in the individual history.

But not only is this order found in the evolution of the whole vertebrate department, but something of the same kind is found also in the evolution of *each class*. The earliest reptiles, the earliest birds, and the earliest mammals had smaller and less perfectly organized brains than their nearest congeners of the present day. This is shown in the accompanying figures (Figs. 57 and 58). To carry out one example more perfectly:

In the history of the horse family, in connection with the changes of skeletal structure already described (page 108), we have also corresponding changes in the size and structure of the brain ; *pari passu* with the improvement of the mechanism we have also increased engine-power and increased muscular energy and therefore increased activity and grace. The brain of a modern horse, though not very large, is remarkable for the complexity of its convolutions.

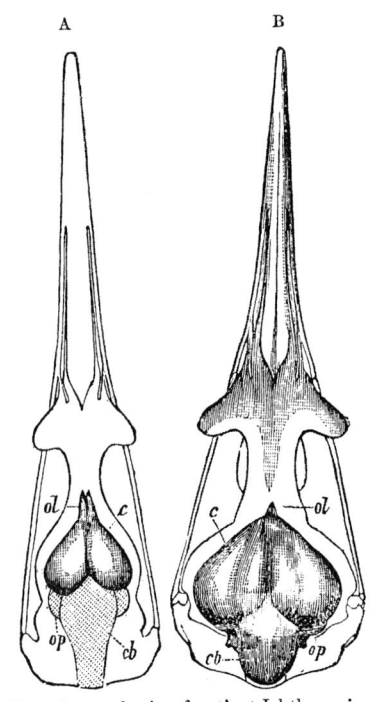

FIG. 57.—A, brain of extinct Ichthyornis ; B, modern tern.

The great energy, activity, and nervous excitability of the horse are the result of this structure.

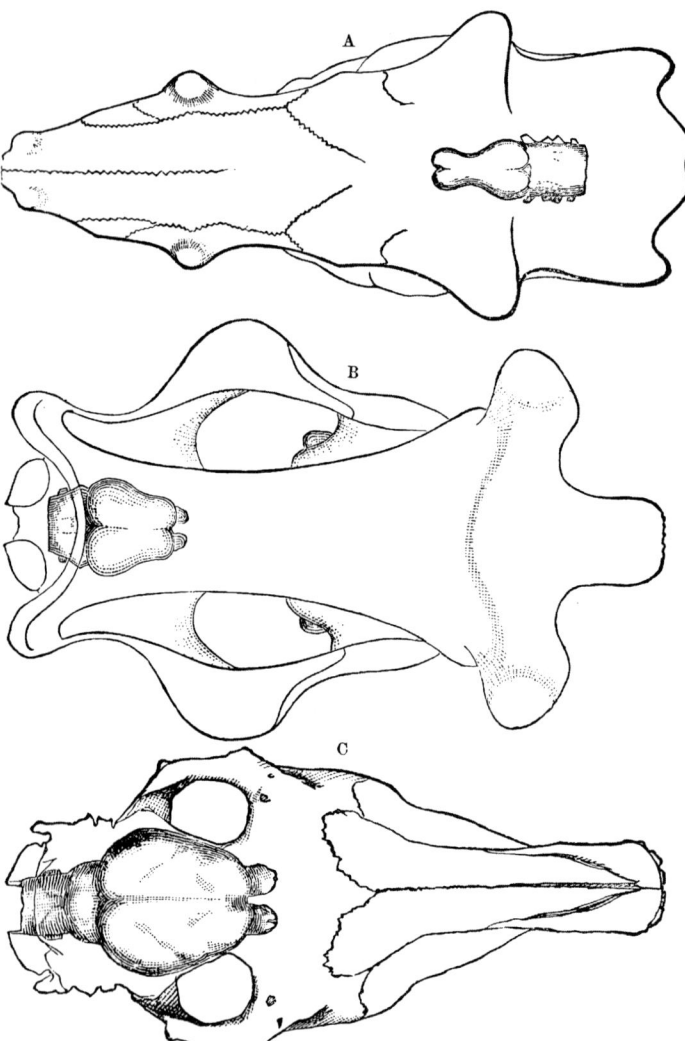

FIG. 58.—A, brain of Eocene dinoceras; B, Miocene brontothere; C, modern horse.

Cephalization.—Thus, in going up the phylogenic, the taxonomic, or the ontogenic series, we find a gradual process of development headward, brainward, cerebrumward ; or, more generally, we might say that in all organic evolution we find an increasing dominance of the higher over the lower, and of the highest over all. For example, in the lowest plane of either series we find first the different systems imperfectly or not at all differentiated. Then, as differentiation of these progress, we find an increased dominance of the highest system—the *nervous system;* then in the nervous system, the increasing dominance of its highest part—the *brain;* then in the brain the increasing dominance of its highest ganglion—the *cerebrum;* and, lastly, in the cerebrum the increasing dominance of its highest substance—the exterior gray matter—as shown by the increasing number and depth of the convolutions. This whole process may be called *cephalization.*

Shall the process stop here ? When evolution is transferred from the animal to the human plane, from the physiological to the psychical, from the involuntary and necessary to the voluntary and free, shall not the same law hold good ? Yes! all social evolution, all culture, all education, whether of the race or the individual, must follow the same law. All *psychical advance is a cephalization*—i. e., an increasing dominance of the higher over the lower and of the highest over all ; of the mind over the body, and in the mind of the higher faculties over the lower ; and, finally, the

13

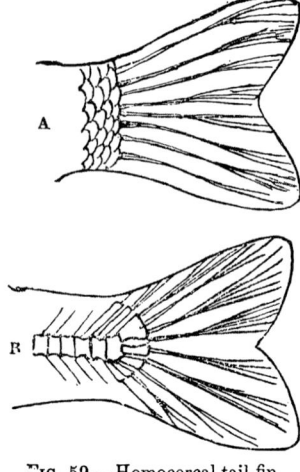

FIG. 59.—Homocercal tail-fin.
A, form ; B, structure.

subordination of the whole to the highest moral purpose.

4. Fish-Tails. — Still another and last example : It has long been noticed that there are among fishes two styles of tail-fins. These are the even-lobed, or homocercal (Fig. 59), and the uneven-lobed, or heterocercal (Fig. 60). The one is characteristic of ordinary fishes (teleosts), the other of sharks and some other orders. In *structure* the difference is even more fundamental than in *form*. In the former style the backbone stops abruptly in a series of short, enlarged joints, and thence sends off rays to form the tail-fin

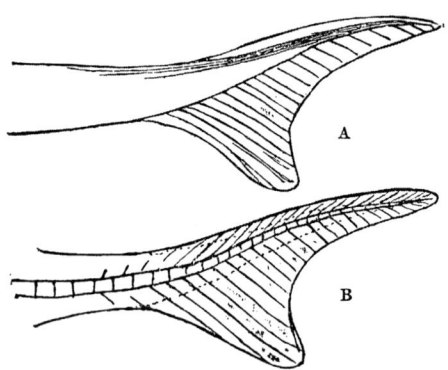

FIG. 60.—Heterocercal or vertebrated tail-fin. A, form ; B, structure.

(Fig. 59, B) ; in the latter the backbone runs through the fin to its very point, growing slenderer by degrees, and giving off rays above and below from each joint, but the rays on the lower side are much longer (Fig. 60, B). This style of fin is, therefore, *vertebrated*, the other *non-vertebrated*. Figs. 59 and 60 show these two styles in form and structure. But there is still another style found only in the lowest and most generalized forms of fishes. In these the tail-fin is vertebrated and yet symmetrical. This style is shown in Fig. 61, A and B.

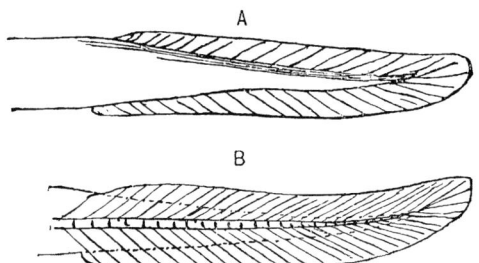

FIG. 61.—Vertebrated but symmetrical fin. A, form ; B, structure.

Now, in the development of a teleost fish (Fig. 58), as has been shown by Alexander Agassiz,* the tail-fin is first like Fig. 61 ; then becomes heterocercal, like Fig. 60 ; and, finally, becomes homocercal like Fig. 59. Why so ? Not because there is any special advantage in this succession of forms ; for the changes take place either in the egg or else in very early embryonic states. The answer is found in the fact that *this is the order of change*

* "Proceedings of American Academy of Arts and Sciences," vol. xiv, May, 1878.

in the phylogenic series. The earliest fish-tails were either like Fig. 61 or Fig. 60 ; never like Fig. 59. The earliest of all were almost certainly like Fig. 61 ; then they became like Fig. 60 ; and, finally, only much later in geological history (Jurassic or Cretaceous), they became like Fig. 59. This order of change is still retained in the embryonic development of the last introduced and most specialized order of existing fishes. The family history is repeated in the individual history.

Similar changes have taken place in the form and structure of birds' tails. The earliest bird known—the jurassic Archæopteryx — had a long reptilian tail of twenty-one joints, each joint bearing a feather on each side, right and left (Fig. 62). In the typical modern

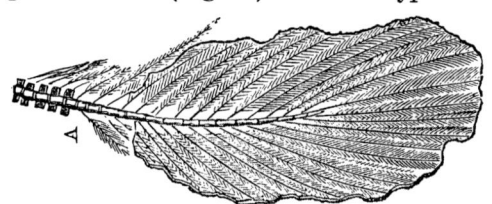

FIG. 62.—Tail of the Archæopteryx.

bird, on the contrary, the tail-joints are diminished in number, shortened up, and enlarged, and give out long feathers, fan-like, to form the so-called tail (Fig. 63). The Archæopteryx' tail is *vertebrated,* the typical bird's *non-vertebrated.* This shortening up of the tail did not take place at once, but gradually. The Cretaceous birds, intermediate in time, had tails intermediate in structure. The Hesperornis of Marsh had twelve joints. At first—in Jurassic—the tail is fully a half of the whole vertebral column. It then gradually shortens up until it becomes the aborted organ of typical modern birds. Now, in embryonic development, the tail of the modern typical bird *passes through all these stages.* At first the tail is nearly one

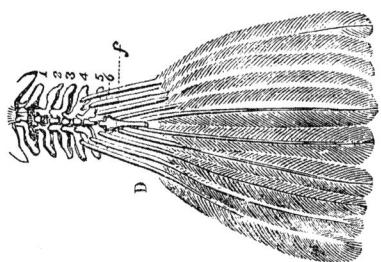

FIG. 63.—Tail of a modern bird.

half the whole vertebral column ; then, as development goes on, while the rest of the body grows, the growth of the tail stops, and thus finally becomes the aborted organ we now find. The ontogeny still passes through the stages of the phylogeny. The same is true of all tailless animals. The frog is tailed in the larval condition, because its ancestors were tailed amphibians. Even man himself is endowed with a much more considerable tail, viz., eight or nine joints, in his early embryonic condition.*

We have taken all our examples from vertebrates, but quite as many and as good examples might be found among articulates. Insects, in the larval state, are worm-like in form. Hence it is probable that the earliest progenitors of this class were worm-like. Again, some insects have aquatic larvæ. The progenitors of

---

* Fol., " Archives des Sciences," vol. xiv, p. 84, 1885 ; " Science," vol. vi, p. 92, 1885.

these—in fact, of all insects—were probably aquatic. Crabs, in a larval condition, are long-tailed, and we know that the long-tailed crustaceans (Macrourans) preceded the short-tailed (Brachyourans). Water-breathing animals preceded air-breathers ; the same is true in the ontogeny of the frog, of many insects, and, we might add, even of mammals. For the breathing of the *fœtus in utero* is essentially by exposure of fœtal blood to the oxygenated blood of the mother in a sort of *gill-fringes* (placental tufts). But why should we multiply examples ? The whole of embryology, in every department, is made up of examples of the same law.

Illustration of the Differentiation of the Whole Animal Kingdom.—Finally, the law of differentiation in the evolution of the whole animal kingdom may be well illustrated by means of the different directions taken in the development of the eggs of all the various kinds of animals. Suppose, then, we have one thousand eggs, representing all the different departments, classes, orders, families, etc., of animals. Many of these may doubtless be identified by form or size, or some other superficial character, as the eggs of this or that animal, *but structurally they are all alike*. At first, i. e., as germ-cells, they all represent the *earliest condition* of life on the earth, and the *lowest forms* of life *now*. If we now watch their development, we find that some remain in this first condition without further change. These we set aside. They are *Protozoa*. The remainder continue to develop, but at first it would be impossible to say to which of the several departments or primary groups they each belonged. Then, by cell-multiplication, the original single cell becomes a cell-aggregate. It may be compared now to a compound protozoan, such as Foraminifera. The cell-aggregate then differentiates into layers, and forms, in fact, a two-layered sac called a gastrula. This is the structure of some of the lowest cœlenterates, such as the hydra. Thus far all seem to go together. But now, for the first time, the primary groups are declared. If it be a vertebrate, for example, the most fundamental characters — the cerebro-spinal axis, the vertebral column, and the double cavity, neural and visceral, are outlined. Suppose, now, we set aside all other departments, and fix our attention on the vertebrates. At first we could not tell which were mammals, birds, reptiles, or fishes ; but after a while the classes are declared. We now set aside all other classes and watch the mammals. After a while the order declares itself. We select the ungulates. Then the family is declared, say the *Equidæ* ; then the genus, *Equus* ; and, lastly, the species, *Caballus*.*

The same would be true if we followed any other line of development, whether in vertebrates or in any other department. Observe, then, that, in following any one line as we have done, there is an increasing speciali-

---

* Of course, this is a purely imaginary case. The conditions of development of the eggs of higher animals forbid continuous watching the process. Yet we do observe in different individuals all these stages in mammals as well as other animals.

zation, and, if we followed all the lines, an increasing differentiation, like the branching and rebranching of a tree. Now, this is the type and illustration of what took place in the development of the animal kingdom. We conclude that the animal kingdom appeared first as Protozoa, then as living cell-aggregates or compound protozoans, then as gastrula or two-layered sacs with oral opening. Then the great primary departments, unless we except the vertebrates, commenced to separate. This took place before the primordial period ; for in the primordial fauna we have all the departments, except vertebrates, already declared. This completely explains why it is that we are able to trace homology only within the limits of each primary group.

But the question has doubtless already occurred to the thoughtful reader, "Why should the steps of the phylogeny be repeated in the ontogeny ?" The general answer is doubtless to be found in the law of heredity— that wonderful law, so characteristic of living things. We have compared it to a brief recapitulation from memory — the minor points, especially if they be also early, dropping out. But can we not explain it further ? It is probable that we find a more special explanation in *"the law of acceleration,"* first brought forward by Prof. Cope. By the law of heredity each generation repeats the form and structure of the previous, and in the order in which they successively appeared. But there is a tendency for each successively-appearing character to appear a little earlier in each successive generation ; and by this means time is left over for the introduction of still higher *new* characters. Thus, characters which were once adult are pushed back to the young, and then still back to the embryo, and thus place and time are made for each generation to push on still higher. The law of acceleration is a sort of young-Americanism in the animal kingdom. If our boys acquire knowledge and character similar to that of adults of a few generations back, they will have time while still young and plastic to press forward to still higher planes.

Proofs from Rudimentary and Useless Organs.—These have to a large extent been anticipated under previous heads. The tails of birds and the gill-arches of reptiles are rudimentary. The finger-bones of a whale's paddle or a turtle's flipper may be regarded as useless, at least so far as the exact number of constituent pieces is concerned ; for an extended surface, without visible joints or separate fingers, is all that is seen, and apparently all that is required. The splint-bones of a horse's foot or the dew-claws of a dog's foot are certainly useless. We have already, in speaking of modifications of structure and of embryonic conditions, given many examples of this kind, but it may be well to add some striking examples with this special point in view.

If different orders of existing mammals were indeed made by gradual modification of some generalized primal form, then it is evident that these useless remnants of once useful parts would be most common in the most highly modified forms. Now, of all mammals, the whales are perhaps the most modified or changed from the original mammalian form — so much modified, in fact, that the popular eye scarcely recognizes them as mammals at all. Here, then, we might expect, and do indeed find, many examples :

1. The baleen whales have no teeth, and no use for them. They have instead a wonderful armature of fringed whalebone plates (baleen), by means of which they gather their food.* Yet the embryo of the whale has a full set of rudimentary teeth deeply buried in the jawbone, and formed in the usual way characteristic of mammalian teeth—i. e., by an infolding of the epithelial surface of the gum—*but the teeth are never cut ;* in fact, they reach their highest development in mid-embryonic life, and are again absorbed. Why, then, this waste of developmental energy ? Why should teeth be formed only to be reabsorbed without being cut ? The only conceivable answer is, because the ancestors of the whale, before the family of whales was fairly established, had teeth which were gradually, from generation to generation, aborted, because no longer used, the baleen plates having taken their place. If whales were made at once out of hand as we now see them, is it conceivable that these useless teeth would have been given them ?

2. Again, many whales have rudimentary pelvic bones, but no hind-limbs. Why should there be pelvic bones,

---

* These baleen plates are not modifications of teeth, as might at first be supposed, but rather of the transverse gum-ridges found on the roof of the mouth of many mammals, and conspicuous in the horse.

when the sole object of these bones is to act as a basis for hind-limbs ? In some whales, for example the right whale, there are also rudiments of hind-legs, but these are buried beneath the skin and flesh, and therefore, of course, wholly useless. The only explanation of these facts is that the ancestors of all the whales before they had become whales were quadrupeds, which afterward took to the water, and little by little the hind-legs, for want of use, dwindled away to the useless remnants which we now find.

3. Again, whales seem to be hairless, yet rudimentary hairs are found in the skin. Their organs of smell are rudimentary, but made on the pattern of those of mammals, not of fishes—i. e., they are air-smelling, not water-smelling organs. From all these, as well as many other facts, it is evident that the whales descended in early Tertiary times from some marsh-loving, powerful-tailed, short-legged, scant-haired quadruped by modifications gradually induced by increasing aquatic habits.

Examples of such rudimentary organs might be multiplied without limit. As might be expected, some are found even in man. Such, for example, are the muscles for moving the ear, necessary in animals but useless in man, and therefore rudimentary. Similarly useless in man are the scalp-muscle, used by aminals to erect the crest or bristles on the head, and the skin-muscle of the neck and chest, used by animals for shaking the skin of those parts. Most persons have lost the power of using these. For my part I can use them all—ear-muscles, scalp-muscle, skin-muscle—but they serve no useful purpose.

Again, and finally, in man and many mammals we find a slender, worm-like appendage about three inches long, attached to the cæcum of the large intestine. Anatomists and physiologists, under the influence of that philosophy which maintains that every part of the fearfully and wonderfully made human frame was *directly* contrived to subserve some useful purpose, have puzzled themselves to find the use of this. It probably has no use; on the contrary, it is a continual source of danger. If the human body had been made at once out of hand, it would not have been there. How came it, then? It is the rudimentary remnant of an organ—a greatly enlarged cæcum—which has served, and in some mammals still serves, a useful purpose. All these cases are survivals; they are organs which, like many customs in society, have outlived their usefulness, but still continue by heredity.

But why multiply examples? All along the track of evolution organs become useless by changes in the habits of their possessors. They are not, however, shed or dropped bodily at once. No; they are *retained by heredity*, but *dwindle by disuse*, more and more, until they pass away entirely. But even when they are entirely gone in the adult, they are often found still lingering in the embryo. They are among the most obvious and convincing proofs of the origin of organic forms by derivation.

CHAPTER VIII.

IT is well known that the kinds of organisms found in widely-separated countries differ more or less conspicuously. The traveler in Australia or in Africa finds all, the traveler in Europe nearly all, the animals and plants wholly different from those he has been accustomed to see at home. Even the visitor from the Atlantic to the Pacific coast, if he observes at all, will find nearly all organisms strange to him. The facts of geographical diversity of organisms are so numerous and complex that, at first sight, they seem utterly lawless. Only recently this subject has been redeemed from chaos and reduced to something like order and law by the light thrown upon it by the theory of evolution. We will give, in very brief outline, the most important facts, and then show how they may be ex' plained.

Geographical Faunas and Floras.—The group of animals and plants inhabiting any locality, whether peculiar to that locality or not, is called, in popular language, its fauna and flora. But, in a true scientific sense, a fauna and flora is a *natural* group of animals and plants in one place, *differing* more or less conspicuously from other groups in other places, and *separated from them by physico-geographical boundaries, or by physical conditions of some kind.* The members of such a group can only exist in certain harmonic relations with external conditions, and with one another. These relations with one another are often complex and nicely adjusted, so that change in one term is propagated through the whole series of terms, giving rise often to the most unexpected results, until finally a new equilibrium is established. Thus, the destruction of certain insectivorous birds, in mere wanton sport, may give rise to the multiplication of insect pests, and this to the destruction of certain kinds of plants, and this to the diminution of certain herbivores, and this in its turn to the disappearance of certain carnivores. It is well known that the introduction of rabbits into New Zealand and Australia has produced the most unexpectedly disastrous effect upon certain crops, on account of the absence of the fierce and active carnivores which keep in check their excessive multiplication in Europe.

Now, among the physical conditions which limit faunas and floras, and separate them from each other, the most important and universal is temperature.

Temperature-Regions.—If we travel from equator to pole, we pass through mean temperatures varying from

80° to 0°. This gives rise to a very regular zonal arrangement of plant-forms : 1. We have first a region in which palms and palm-like forms are abundant and characteristic, and which therefore may be called the region of palms. It corresponds with the tropic zone. 2. We next have a region in which hard-wood foliferous trees are most abundant and characteristic ; first mostly evergreens and then deciduous trees, and therefore may be called the region of hard-wood forests. This corresponds with the temperate-zone. 3. Then we find a region characterized predominantly by pines and pine-like trees and birches, and may be called the region of pines. This is the sub-Arctic region. 4. Then a region without trees, but only shrubs and summer plants. This is the Arctic region. 5. And, finally, an almost wholly plantless region of perpetual ice—the polar region.

These regions are determined wholly by temperature, and therefore, in going up a mountain-slope to snowy summits, we pass through similar regions in smaller space. For example, in going from sea-level to the summits of the Sierra, 14,000 to 15,000 feet high, we commence in a region of predominantly hard-wood trees ; but at 3,000 feet the forests become almost wholly coniferous, at 11,000 to 12,000 feet the vegetation becomes shrubby, and at 13,000 feet we reach perpetual snow.

We have taken plants first, because these, being fixed to the soil and incapable of voluntary seasonal migrations, are more strictly and simply limited by temperature—i. e., the arrangement of different kinds in zones is more simple and conspicuous. But the same rule holds also for animals. In passing from equator to pole, animal kinds also change frequently, so that there are many temperature-faunas in which the animals are all very different. In both animals and plants, species, genera, families, etc., are limited by temperature. These are familiar facts ; we recall them to the reader in order that we may base thereon a clearer definition of these limits.

More Perfect Definition of Regions.—1. The area over which any form spreads is called its *range*. Now, the range of a species is more restricted than that of a genus, because, when a species is limited by tempera- ture, another species of the same genus may carry on the genus. For the same reason the range of a family is usually greater than that of a genus, and so on for higher classification-groups. For example, pines range on the slopes of the Sierra from about 2,000 feet to 11,000 feet, but not the same species. In ascending, we meet first the nut-pine (*Pinus Sabiniana*), then the yellow-pine (*P. ponderosa*), then the sugar-pine (*P. Lambertiana*), then the tamarack-pine (*P. contorta*), and last, the *Pinus flexilis*, etc.

2. Where two contiguous temperature-regions come in contact, there is no sharp line between ; on the con- trary, they *shade gradually*, almost imperceptibly, into one another, the ranges of species overlapping and in-

terpenetrating, and the two species coexisting on the borders of their ranges. This is represented by the diagram (Fig. 64), in which the horizontal lines represent the north and south ranges of species of two groups, A and B, separated by the dotted line.

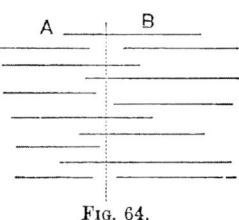

FIG. 64.

3. Species also pass out gradually on the borders of these ranges and others come in gradually, so *far as number and vigor of individuals are concerned.* If *a a'* and *b b'* (Fig. 65) represent the north and south range of two species, and *b a'* their overlap or area of coexistence, then the height of the curves A and B will represent the

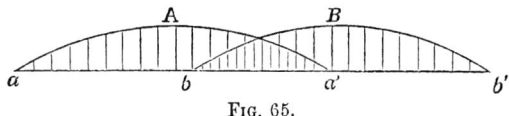

FIG. 65.

number and vigor of the individuals in different parts of the range.

4. While, therefore, there is a shading of contiguous groups into each other by overlap of species-ranges; while there is also a gradual passing out of species so far as number and vigor of individuals is concerned, yet, in *specific characters* we observe usually no such gradation. Species seem to come in on one border with all their specific characters perfect, remain substantially unchanged throughout their range, and pass out on the other border, still the same species. In other words, one species takes

14

the place of another, usually by *substitution*, not by *transmutation*. It is *as if* species had originated, no matter how, each in its own region, and had spread in all directions as far as physical conditions and struggle with other species would allow. This important subject will be more fully discussed later.

5. We have thus far spoken of species as limited by temperature alone, but they are limited also by *barriers*. If, then, there be an east and west barrier, such as a high mountain-range, or a wide sea or desert, there will be no shading or gradation of any kind, because the barrier prevents overlapping, interpenetration, and struggle on the margins. For example : The species north and south of the Himalayas, or north and south of Sahara, are widely different. It is, again, *as if* they originated each where we find them and spread as far as they could, but the physical barrier prevented mingling and shading.

6. There are temperature-regions south as well as north of the equator. Now, although the climatic conditions are quite similar, the species of corresponding temperature-regions north and south are wholly different. It is, again, as if they originated where we find them, and were kept separate by the barrier of tropical heat between. If carried over, they often do perfectly well.

Continental Faunas and Floras.

If the land-surfaces were continuous all around the globe, there is little doubt that each temperature region with its characteristic species would also be substantially

continuous. There would, it is true, be some local variations dependent upon soil and humidity, etc., but substantially the same species would exist all around. The
distribution would be almost wholly zonal. But the in-
tervening oceans are complete barriers to continental
species. Hence we ought to expect, and do find, that the
faunas and floras of different continents are almost totally different.
Each apparently
originated on its
own continent,
and did not
spread to other
continents, only
because they
could not get
there. It is ne-
cessary to explain
this in more de-
tail.

Fig. 66 repre-
sents a polar view

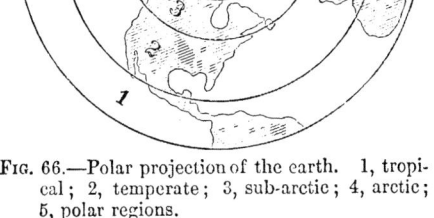

FIG. 66.—Polar projection of the earth. 1, tropical; 2, temperate; 3, sub-arctic; 4, arctic;
5, polar regions.

of the earth, showing the eastern and western continents, and the five temperature zones already described.
Now, if we examine the species in each region, commencing at the pole, we find that those of Nos. 5
and 4 are almost identical all around. The reason is
obvious. The continents come close together there,
with ice-connection if not land-connection all around.

There is but one circumpolar region. But, as soon as we come down to No. 3 and No. 2, the species on the two continents are nearly all different, because there is an impassable barrier between, either in the form of ocean or of Arctic cold. For example, the animals and plants inhabiting the United States are almost wholly different from those in Europe, not only in species, but even largely in genera and to some extent in families. There are some exceptions to this rule, but these are of the kind which prove the rule, or rather the principle on which the rule is founded. These exceptions are mainly of three kinds : 1. *Introduced species.*—All our weeds, many garden-plants, and many animal pests are of this kind. They were not found here when America was discovered, only because they could not get here ; for, when brought here, they do so well that they often overrun the country and dispossess the native species, as we ourselves have done the Indians. 2. *Hardy or else wide-migrating species.*—Hardy species have wide range ; they may belong to No. 4 as well as No. 3. If so, they range down to No. 3 on both continents. Migrating birds, such as ducks and geese, etc., breed in summer in No. 4, and migrate southward in winter on both continents from the common circumpolar ground. 3. *Alpine species.*—It is a curious fact that species on tops of snowy mountains in temperate regions of the two continents are wonderfully similar, though so completely isolated. We are not yet prepared to discuss this point. We shall do so later. Suffice it to say now that it can be completely explained.

In region No. 1 the continental diversity is still greater. Not only species and genera, but whole families and even orders, are peculiar to each continent. The great pachyderms—elephant, rhinoceros, hippopotamus— are peculiar to the Eastern ; the edentates—sloths and armadillos—to the Western. The humming-birds, those gems of the forests, of which there are over four hundred species, and the whole cactus family, are peculiar to America, while the tailless monkeys are equally characteristic of the Eastern Continent.

The continents do not come together again toward the south, and, therefore, as might be expected, the great difference between the two persists to the southern points. The faunas of the southern points of South America, Africa, and Australia are very different.

Subdivisions of Continental Faunas and Floras.—Besides the subdivisions of continental faunas, north and south, determined by temperature as already explained, if there be in any continent an impassable barrier running north and south, there will be a corresponding difference in the species on the two sides, east and west. We give but one example : The North American Cordilleras or Rocky Mountains, with their high ranges and desert plains, constitute a very great barrier between the eastern and western portions of the United States. Hence, we find an extraordinary difference between the species inhabiting California and those found in the eastern portion of the country. Speaking generally, all the species and many of the genera are peculiar. The exceptions, too, are significant. Leaving out introduced species, of which there are many, they are mostly strong-winged or widely-migrating birds, such as the turtle-dove, the turkey - buzzard, the bald eagle, and, of course, many water-birds.

Special Cases.—If any body of land is widely separated from all other lands by deep seas, we invariably find a corresponding peculiarity of its species. Thus, the species inhabiting Australia and Madagascar are perhaps the most peculiar in the world. We do not dwell further on these, because we will discuss them hereafter. There is a little group of very small islands—the Galapagos—about six hundred miles off the western coast of South America, and surrounded on all sides by deep sea. These islands are stocked with a collection of curious animals not found elsewhere on the surface of the earth ; but among them are no mammals at all. We might multiply examples without limit. Even the rivers emptying in the same sea sometimes have each its peculiar species of mussels. In the Altamaha River there are several species of unios—such, for instance, as the wonderful spinous unio—not found elsewhere. How came they there ? Howsoever they may have come there, they are now kept isolated there by barriers of land and of salt water.

Many other curious details will come up in our discussion of the origin of diversity.

Marine Species.—Precisely the same principles apply here ; but diversity in the case of marine species is perhaps less marked, and certainly less general, because of the universal oceanic connection. Open-sea species are therefore almost universal. But many marine species are confined to shallow water, and therefore to shore-lines. The species on the two shores of the same ocean, or the two coasts of the same continent, are different, being isolated east and west by barriers of deep sea or of land, and north and south by temperature. Also about isolated lands, like Australia and Madagascar, the species are peculiar.

Thus, then, species, genera, etc., are limited in every direction ; north and south by temperature, and in all directions by barriers, in the form of oceans, deserts, and mountain-chains. Add to these, peculiar climates and soils, and we see that, from this point of view, the whole surface of the earth may be divided and subdivided into regions, sub-regions, provinces, etc. It would carry us too far to explain the primary and secondary divisions adopted by Mr. Wallace, and the somewhat different ones suggested by Mr. Allen. Our main object is to discuss the *cause* of this diversity, and especially to show the light shed upon it by the theory of evolution. We have only given a sketch of the facts sufficient for this purpose.

Theory of the Origin of Geographical Diversity.

It will be observed that all along we have assumed a sort of provisional theory. We have said in every case, it is *as if* organic forms originated where we find them, and have gone thence wherever they could—as far in every direction as physical conditions and struggle with competing species would allow. This view has been formulated as the "theory of specific centers of origin." There would be less objection to this as a first provisional theory did it not assume a supernatural *mode* of origin. But, in the minds of those who hold it, it has usually assumed expressly or tacitly the form of *"specific centers of creation,"* thus implying the immutability of specific types and the supernaturalism of specific origin (page 68). In this latter or usual form it completely fails to account for the facts given above. For, if this were the mode of origin, each species ought in every case to be perfectly adapted to its own environment, and to no other. But, on the contrary, introduced species often flourish better than in their own country, and better than the natives of their new homes. In the less objectionable form of "specific centers of origin," without defining the mode of origin, it accounts well for many of the more obvious facts of geographical diversity, as it *now* exists, but not all. According to this view, the amount of diversity ought to be in strict proportion to the completeness of isolation, or impassableness of the separating barriers; but this is not *exactly* true. There is another element, not yet mentioned, which is just as important as impassableness, but which until recently has been left entirely out of account. This is the element of *time*—the amount of time since the barrier was set up, or during which it has continued to exist. These two elements, it is true, are closely connected with each other ; for, since all changes in physical geography have taken place very slowly—since barriers in the form of mountain-ranges and seas have increased by slow process of growth—it is evident that impassableness is, to some extent, a measure of time. But they are by no means in strict proportion. The one or the other may predominate.

Now, this time-element connects geographical distribution with changes of physical geography and climate in *geological* times, and especially with the *latest* of these changes, viz., those occurring during the *Glacial epoch*. During that remarkable epoch extraordinary changes of climate, from extreme Arctic rigor to great mildness, enforced wide migrations of species southward and northward ; while concomitant changes of physical geography, by elevation of the earth's crust over wide areas, opened highways between previously-isolated continents, permitting migrations in various directions, and by subsequent depression again isolating the migrated species in their new homes. It is evident, then, that the recognition of the element of almost unlimited time at once introduces into the question of geographical distribution the *idea of evolution*. If the study of geographical distribution, as *it now exists*, and as a part of science of physical geography, gave rise naturally to the theory of "specific centers of origin," the study of the same, in connection with geological time, and as a part of geological science, now demands its explanation by the theory of evolution.

It must be borne in mind, then, that geographical diversity of organisms is not a question of the present epoch only. There has been geographical diversity in every previous geological epoch ; it is, therefore, a question of geology as well as of biology. It is probable, however, that diversity has increased with the course of geological times, and is greater now than ever before. In other words, in the evolution of the organic kingdom, the law of differentiation has prevailed here, as in other departments of biology. A clear statement of the causes of the *present* distribution of organisms must embrace also the causes of geographical diversity *generally*. We give, therefore, at once a brief statement of what seems to us the most probable view, and shall then proceed to show how it explains the present distribution.

Most Probable View of the General Process.—Bearing in mind, then, this time-element, the phenomena of geographical diversity are best explained by the following suppositions : 1. A gradual progressive movement (evolution) of the organic kingdom, marching, as it were, abreast, at equal rate along the whole line— i. e., in all parts of the earth, and throughout all geological times, under the action of all the forces or factors, and following all the laws, of evolution already explained (pages 19 and 73). If this were all, there would be no *geographical* diversity, although *organic diversity*

might be as great as it is now. There would be differentiation of forms and structure everywhere, but no
differentiation of groups in different localities. 2.
Under the influence of different conditions in different
places, more or less isolated from one another by climatic or physical barriers, the onward movement (evolution) of organic forms takes different directions and
different rates, and gives rise to local groups, which
become more and more differentiated, without limit as
time goes on. This element, acting by itself throughout all geological times, would ere this have produced
an extreme geographical diversity, such as does not anywhere exist. 3. From time to time, at long intervals,
extensive changes of physical geography and climate,
produced by crust elevations, partly enforce by change
of temperature, and partly permit by opening of gateways, extensive migrations and dispersals of species, by
which mingling and struggle for life and final readjustment takes place, and extreme diversity is prevented.
Such mingling of different faunas and floras on the same
ground, and the severe struggle for life that thus ensues,
and the survival of the fittest in many directions, are, as
already shown, among the most powerful factors of evolution. They tend to *increase organic* diversity, but to
diminish geographical diversity. 4. At the close of such
great periods of change as indicated in the last, by contrary movement of the earth-crust—i. e., subsidence—
new barriers are set up and new isolations are produced,
and the process of divergence again commences and

increases steadily so long as the barriers continue to exist.

Now, the last of these periods of great changes and extensive migrations, and subsequent isolations, was the Glacial epoch. It was this epoch, therefore, which mainly determined the present geographical distribution of species. Thus, the present distribution is a key to the directions of the last great migrations, and therefore to the nature of the changes in physical geography and climate which then occurred ; and, conversely, the character of these changes, determined in other ways, *furnishes the only key to the present distribution of species.*

Before applying the foregoing principles in the explanation of special cases, it may be well to give a very brief outline of the condition of things during the Glacial epoch.

In America, during this epoch, by increasing cold the southern margin of the great northern ice-sheet crept slowly southward, until it reached the latitude of about 38° to 40°. Arctic species were thus driven southward slowly, from generation to generation, until they occupied the whole of the United States, as far as the shores of the Gulf, while temperate species were forced still farther south, into Central and South America. This period of extreme rigor and southward migration was followed by a period of great mildness, during which the ice and its accompanying Arctic conditions retreated northward, followed by Arctic species.

More than one advance and retreat, apparently, occurred during this time. Again, during the same time,
brought about by northern elevation, there was broader
connection than now exists between North and South
America, and free migrations between, in both directions, enforced by extreme changes in temperature.
Also, during this or previous time, there were broad
connections between North America and Asia, in the
region of Behring Strait, and between America and
Europe, in high-latitude regions, and extensive migrations of faunas and floras between were thus permitted.
The necessary result of all these migrations of species,
partly enforced by changes of climate, partly permitted
by opening of gateways since closed, was exceptionally
rapid changes in organic forms. This was the result of
two causes : First, the severer pressure of a changing
physical environment ; and, second, a severer struggle
for life between the natives and the invaders.

In Europe, during the same time and from similar
causes, there were at least three or four different faunas
struggling together for mastery on the same soil. First,
there were the Pliocene indigenes, who had, if any,
pre-emption right to the soil ; second, invaders from
Arctic regions, driven southward by increasing cold ;
third, invaders from Asia, permitted by the removal
of the old sea-barrier which once extended from the
Black Sea to the Arctic, and of which the Caspian and
Aral are existing remnants, and thus opening a gateway
for migration which has remained open ever since ;

fourth, invaders from Europe and Asia into Africa, and sometimes back again into Europe, by opening of gateways through the Mediterranean, which have been since closed. One of these highways was through Gibraltar, and one from Italy to Africa through Sicily. As in America, so here, in even greater degree, the severe pressure of changing environment and the severe struggle for life produced rapid changes of organic forms. Many species were destroyed; others saved themselves by modifications adapted more perfectly to the changed conditions. There is little doubt that man came into Europe with the Asiatic invasion, and was one of the principal agents of change, especially in the way of destruction of many old forms.

Such is a very brief outline of the last great geological change and its general results. Being the last, this one has left the strongest and most universal impress on the *present* geographical distribution. But similar changes by crust oscillations, if not also by extreme changes of climate, have repeatedly occurred in geological times, and some of the most remarkable geographical faunas and floras are the result of these earlier geological changes. We will now give a few examples illustrating these principles:

1. *Australia* is undoubtedly more peculiar in its fauna and flora than any other known country. Not only are all its species peculiar, not found elsewhere on the face of the earth, but its genera, its families, and even many of its orders of animals and plants, are

also peculiar. These facts are so familiar that it is unnecessary to dwell on them. I need only mention, among plants, the whole of the simple-leaved acacias, already mentioned on page 86, of which there are so many species, and the whole family of the eucalyptids, of which there are several hundred species. Among animals I need mention only the order of monotremes, or egg-laying mammals, and nearly the whole order of marsupials, or pouched animals, of which there are over two hundred species. On the other hand, the true typical mammals are entirely absent, with the exception of a few bats and a few rats, which have evidently been accidentally introduced from abroad.

Another very noteworthy fact, which must be taken in connection with the last, is that Australian forms are far less advanced in the race of evolution than those of any other country—i. e., that many old forms which have long ago become extinct elsewhere are still retained there. A few examples will suffice. The marsupials just mentioned are an old form once universally distributed, but now nearly extinct everywhere, except in Australia ; the cestracion, or Port Jackson shark, and the ceratodus, are Palæozoic and Mesozoic forms retained only in Australia.

What is the explanation of these remarkable facts ? We find the sufficient answer in the fact that Australia has been long isolated from all other countries. While geographical changes in geological times have mingled more or less the organic forms of other countries, and the sharp struggle for life has produced more rapid advance and the production of many new and higher forms better armed for the battle of life, Australia has remained isolated from competition, and therefore comparatively unprogressive.

Can we tell when Australia was finally isolated? Approximately we can. The class of mammals is divided into two groups, which differ widely from each other; so widely, that they are called sub-classes. These are placental mammals, or true typical mammals, and non-placental or reptilian mammals. The non-placentals include only the marsupials and the monotremes (ornithorhyncus and echidna). The monotremes actually lay eggs and incubate them. In the marsupials the embryo has no placental connection with the mother, and is born in a very imperfect condition, utterly unfit for independent life, and placed in the pouch (marsupium), and *permanently* attached there to the teat until it is capable of independent life; after which only it voluntarily nurses like other new-borns. In other words, the gestation commenced in the womb is completed in the pouch. The uterine gestation in the opossum is only seventeen days, while the marsupial gestation is about two and a half months. In a kangaroo seven feet high in sitting position the embryo at birth is only one inch long—a pink, hairless, almost amorphous mass. The monotremes are pure oviparous animals, like birds and reptiles. The marsupials might well be called *semi-oviparous*. In pure egg-layers the whole embryonic de-

velopment is outside of the body ; in pure young-bearers
the whole is within the body ; in marsupials it is partly
within and partly without. Now—1. The monotremes
are found nowhere but in Australia and the neighboring New Guinea. 2. The marsupials are also all confined to the Australian region, except a few oppossums
in America. 3. There are some two hundred and thirty species of non-placentals in the Australian region.
4. As already said, there are no true mammals at all in
Australia, except a few bats and rats which have come
accidentally from abroad. 5. But non-placentals existed
abundantly in *Mesozoic times everywhere*, both in Eu-
rop-Asia and in America, while true mammals did not
appear at all on the surface of the earth until the *Tertiary*, when they almost immediately became very abundant everywhere, except in Australia. *Evidently, therefore, Australia was isolated before the Tertiary.* The
enormous difference between its fauna and flora and
those of other countries is due to at least three things :
1. So long an isolation necessarily produced great divergence of forms. This alone, however, would not affect
the *grade of organization*. 2. Saved from wide migrations, and especially invasions from Eurasia, the great
field of competitive struggle, it was left far behind in
the race of evolution. Hence many of its forms are ar-
chaic ; its mammalian fauna, for instance, is still in the
Mesozoic stage. 3. Its distance from other large continents is so great that accidental colonization has been
very slight, only extending to a few bats and a few rats.

15

I stop a moment to insist on the effect of competitive struggle in developing organic forms strong for the battle of life. Of all the continents, Eurasia has been the scene of most frequent geological changes, and therefore the arena of fiercest competitive struggle through wide and frequent migrations. Eurasian species, therefore, are the strongest of all. They have conquered wherever they have gone. Species in isolated regions are usually the weakest. The great moas and the dodo could not have continued to exist unless protected in a sort of bomb-proof. Kangaroos would now be quickly exterminated by the introduction of fierce Eurasian carnivores.

2. *Africa.*—The fauna of that part of Africa north of Sahara is essentially Mediterranean—i. e., a sub-group of the Eurasian. Sahara, rather than the Mediterranean Sea, is the true intercontinental barrier. The true African region, therefore, is south of Sahara. Now, according to Mr. Wallace, whom I mainly follow here, the true African mammalian fauna consists of two very different groups of animals. The one is a group of very small, curious animals, mostly low forms of insectivores and lemurs, very peculiar to this region, though more resembling those of Madagascar than of any other region ; the other is a group of large and powerful animals which dominate the region. These latter are similar to, though not identical with, those which inhabited Eurasia in Pliocene times. The great carnivores, pachyderms, and ruminants of the region are examples of this group. Now, the explanation of these facts is as follows : The indige-

nes of Africa are the animals of the first group. Africa, in Tertiary times, was isolated from the great field of combat, Eurasia, and therefore its animals were small, of low grade, and peculiar. During later Tertiary (Pliocene) times, then, Africa was inhabited by animals of the first group, while Eurasia was dominated by animals of the second group. These two groups were then separated by the Desert of Sahara, or else by a sea in that region. Some time during the Glacial epoch geographical changes removed this barrier, and climatic changes drove the Eurasian animals southward into Africa, where, finding congenial climate, they took possession of the continent, dominating the feebler natives. Subsequently they were isolated there by the formation of the desert, and the process of divergence commenced, and has gone on to the formation of many new forms. Meanwhile the change, partly by extinction and partly by modification, has gone on still more rapidly in Eurasia, but in a different direction. Hence, Africa is regarded as one of the primary faunal regions.

3. *Madagascar.* — This, next to the Australian, is probably the most peculiar faunal region known. There is probably not a single mammalian species found there which is known to occur anywhere else. It is remarkable also as the principal home of that strange, generalized, ancient form of monkeys—the lemurs. And yet its animals, though very different, have a distant resemblance to those of Africa ; not, however, to the present dominant type, but to those we have called the indigenes. Not

one of the northern invaders is found there. The obvious conclusion from these facts is, that Madagascar was formerly united with Africa, and both were occupied by the same mammalian fauna (which may be called African indigenes, although they were considerably different from their descendants of the present day), but became separated before the northern invasion. The effect of this invasion was to hasten the steps of change in the indigenous fauna of Africa, partly by extermination, partly by modification, while the isolated portion in Madagascar went on at the usual slow rate of change in isolated regions. The time since the separation (which was certainly during the Tertiary period) has been sufficiently long to produce very great divergence in both, but *especially in the African indigenes*. In the fauna of Madagascar, therefore, we have a nearer approach to the original fauna of both. On account of this long isolation, we have here many ancient types which are extinct elsewhere. The lemurs are such an ancient type. These are a wonderfully-generalized type of monkeys—a connecting link between monkeys and other mammals, especially insectivores. As might be supposed, from the law of differentiation, already explained (page 11), they are the earliest form, the progenitors, of monkeys. In fact, in early Tertiary times, they were found not only in Africa and Madagascar, but all over the earth, as the only representatives of the monkey family. The true monkeys were not introduced until the mid-Tertiary. In Eurasia and in America (which at that time was probably connected with Eurasia) wide migrations and frequent conflicts of faunas produced comparatively rapid evolution of new and higher forms, while in isolated Africa old types continued until the invasion. Madagascar was spared this invasion, and therefore old types are still preserved there. At present, at least three quarters of all lemurs are confined to Madagascar, although a few species are still found in Africa and in the great East Indian islands.

4. Island-Life.—Mr. Wallace has divided islands into two kinds, continental and oceanic islands. The division is undoubtedly a good one, although we may not always be able to refer an example with certainty to the one or the other class. *Continental* islands are those on the borders of continents, and separated from the latter only by *shallow water*. *Oceanic* islands are those, usually very small, found in the midst of the ocean, with abyssal depth all about. Continental islands may be regarded as appendages to the neighboring continent—as outliers of continents separated by submergence, and have, in fact, been thus formed. Oceanic islands have been formed geologically recently by volcanic action building up from the sea-bottom. Continental islands have a continental structure—i. e., they are composed of stratified as well as of igneous rocks. Their structure is a record of geological history, like that of the neighboring continent. Oceanic islands are composed wholly of volcanic rocks ; or, if there be any stratified rocks, these are only of the most recent date. As examples of continental islands we

have New Zealand as an appendage of Australia, the great East Indian (Borneo, Java, Sumatra, etc.) and the Japanese Islands, etc., as appendages of Asia ; the British Islands, appendages of Europe ; the West Indian Islands, appendages of America ; Madagascar, an appendage of Africa, etc., etc. As examples of oceanic islands we have the Azores and Bermudas in the Atlantic, and the Polynesian islands in mid-Pacific.

a. Continental Islands.—Now, the fauna of continental islands, as might be expected from the mode of origin of these islands, is similar to, though not identical with, that of the neighboring continent; the amount of difference being in proportion to the *length of time since* they were separated and the *width of the separation.* *Madagascar*, for example, has been long separated from its parent continent, and by a wide and deep channel. Its fauna, therefore, differs greatly from that of Africa, although resembling it more than that of any other country. The separation of *New Zealand* from Australia has been not quite so long, and the divergence, therefore, is not so great. These two will be sufficient illustrative examples of long separation, and therefore of great differentiation of forms.

On the other hand, the British Isles are an excellent example of comparatively recent separation. These isles have probably been several times united and separated from Europe, but we are here concerned only with the more recent. They are now separated from the continent and from one another only by shallow seas. An

elevation of less than six hundred feet—geologically a
very small change—would bare the bottoms of the Irish
and English Channels and the North Sea, and connect

FIG. 67.—Map of outline of coast of Western Europe, if elevated 600
feet (after Lyell).

these islands with one another and with the continent (Fig. 67). Now, it is well known that there were during the Glacial epoch, and subsequently, several oscillations of level sufficient to connect and separate these islands. In the mid-Glacial epoch the British Islands, by submergence, were nearly obliterated, being reduced to an archipelago of small islets representing the high mountains of Wales and Scotland. The Pliocene fauna and flora were, therefore, largely exterminated. During the close of that epoch they were elevated above the present condition and broadly connected with the continent (Fig. 67), and the newly-exposed land was taken possession of by European species, man among the number. Still later—i. e., at the beginning of the present epoch—the islands by subsidence were again separated, but not widely, from the continent. This is the condition now. What, then, was the result? 1. The fauna and flora of the British Isles are substantially the same, but *less rich* in species than that of Continental Europe, some of the European species being wanting. This shows that the last connection was not a long one.; the colonization had not been completed before re-isolation. 2. This poverty of species is more conspicuous in Ireland, because colonization is progressive in space as well as in time. Some species had not reached so far when Ireland was re-isolated from England. The conspicuous absence of snakes, for example, is thus accounted for. There is, we all know, another theory to account for this, but we prefer the natural one. 3. The difference between British and European fauna and flora is very small, it is true, but there is some difference, varietal if not specific. The reason is, that the time since separation is too small to produce much divergence, and the width of the existing barriers not great enough to prevent colonization by accidental causes.

The continental islands of the southern coast of Asia are good examples of an intermediate condition as to the length of time since separation, and of the consequent degree of differentiation of the faunas and floras.

Coast-Islands of California.—We give one more example, and dwell upon it a little, because it occurs on our own coast.

The recent studies of Mr. E. L. Greene on the flora of the islands off the coast of California have brought to light some facts which are an admirable illustration of the principles laid down above.

On looking at a good map of California, any one will observe eight or ten islands, some of them of considerable size, strung along the coast from Point Conception southward, and separated from the mainland by a sound twenty to thirty miles wide. They are in structure true continental islands—outliers of the mainland separated by a subsidence of a few hundred feet. Moreover, the date of their separation is known. They were certainly connected with the mainland during the later Pliocene and early Quaternary, for bones of the mammoth, characteristic of that time, have been found on one of them.* They were therefore separated during the Glacial epoch.

The main peculiarities of the flora of these islands are the following :

1. Out of nearly three hundred species of plants gathered by Mr. Greene, about fifty are wholly peculiar to these islands. 2. Of the remaining two hundred and fifty species, nearly all are distinctively Californian. In other words, the distinctively Californian forms are very abundant, while the common American forms are rare—i. e., the island flora is distinctively Californian, with many peculiar species added.

I explain these facts as follows : The whole coast-region of California is geologically very recent, having emerged from the sea as late as the beginning of the Pliocene epoch. As soon as emerged it was of course colonized from adjacent parts. Since that time its peculiar flora has been formed by gradual modification. The environment has been sufficiently peculiar, the isolation sufficiently complete, and the time sufficiently long, to make a very distinct group of organisms. It is one of Mr. Wallace's primary divisions of the Ne-arctic region.

During late Pliocene and early Quaternary times, as already said, the islands were still a part of the mainland, and the whole was occupied by the same species, viz., the distinctively Californian species now found in both, together, as I suppose, with the peculiar island species.

---

* "Proceedings of the California Academy of Science," vol. v, p. 152. 1873.

During the oscillations of the glacial times the islands were separated by subsidence of the continental margin. Simultaneously with this subsidence, or subsequently thereto, came the invasion of northern species, driven southward by glacial cold. Then came the mingling of invaders with natives, the struggle for mastery, the extermination of many forms—viz., the peculiar island species—and the slight modification of others, and the final result is the California flora of to-day. But the island flora was spared this invasion by isolation. Therefore the invading species are mostly wanting, the distinctive island species were saved, and the result is the island flora of to-day. The island flora, therefore, somewhat nearly represents the Pliocene indigenes of both.

It will be observed that this case is somewhat like that of Madagascar, but with a characteristic difference. In the case of Madagascar, the separation has been long. The extreme peculiarity of its fauna is the result partly of progressive divergence and partly of many forms saved by isolation. In the case of the coast-islands of California, the time has not been long enough for any great divergence by modification. The peculiarity of its species is due almost wholly to species saved by isolation.*

b. Oceanic Islands.—We have seen that faunas and floras of continental islands are somewhat similar to those of the neighboring continent, though with varying degrees

---

* For fuller discussion of this subject, see " Bulletin of the California Academy of Science," No. 8, 1887, and " American Journal of Science," for Dec., 1887.

of difference—the amount of difference, or divergence by evolution, being in proportion to the amount of time and the impassableness of the separating barriers. But ocean-ic islands have never been connected with any continent. They are new land formed in the midst of the ocean by volcanic action. When they first appeared they were, of course, without inhabitants of any kind, animal or vege-tal. How were they peopled? We answer by *waifs* from here and there—by *castaways from other lands*. The dominance of particular kinds will depend on the direction of winds and currents, bringing from some lands more than others, and upon the kinds of animals or seeds of plants most liable to be successfully carried across wide seas. Their faunas and floras, therefore, are characterized by a mixture of species resembling, though not usually identical with, those of various lands, with a predominance of those of some one land, and by the singular and complete absence of mammals and amphib-ians, these being unlikely to be transported by floating timber, as are small reptiles and insects, etc. Among mammals, however, there is a significant exception in favor of bats, the reason being both their power of flight and their habit of concealment in hollow trees, etc. To this explanation, however, we must add that divergence by isolation will meanwhile go on in proportion to time. The Azores, for example, have been peopled from Eu-rope, Africa, and America, but mostly from Europe, on account of the prevailing winds and currents being favor-able to colonization from that direction. There are

many curious peculiarities in the species, however, because colonization is very slow, and divergent variation
has been going on *pari passu*. The Bermudas, on the
other hand, have been colonized mainly from America,
because of the current of the Gulf Stream.

These few examples are sufficient for our purpose,
which is only to illustrate the causes of geographical distribution. If any one desires to pursue this interesting
subject, we would refer him to that most fascinating
book, Mr. Wallace's " Island-Life."

5. **Alpine Species.**—These afford an admirable illustration of the fact that in isolated faunas and floras the
amount of difference is proportioned not only to the
completeness of isolation, but also and mainly to the
time of isolation.

It is well known that Alpine species—i. e., those species inhabiting the region bordering the perpetual snow
of lofty mountains—are very similar to one another, even
in the most distant localities, where their isolation from
one another is as complete as possible ; as, for example, in
the high Alps of Europe, the high mountains of Colo-
rado and California. Why is this ? We find the key to
this mystery in the additional fact that *they are similar
also to Arctic species*. A somewhat full explanation is
here necessary.

During Miocene times, magnolias and taxodiums (bald
cypress), like those in forests and swamps of Carolina
and Louisiana, and sequoias and libocedrus like those
now in California, and many other temperate - region

forms of plants, grew abundantly in Greenland, and northward certainly to 75° north latitude. At that time there could not have been any perpetual polar ice, and therefore no Arctic species, unless on *high mountains in polar regions*. In Pliocene times perpetual polar ice, and therefore Arctic species, probably commenced to appear. As the cold of the Glacial epoch came on and increased in severity, the polar ice extended southward as a general ice-sheet, until it reached in America 40° and in Europe about 50° north latitude. In the United States its margin can be traced as a distinct moraine through Long Island, middle New Jersey, middle Pennsylvania; thence, less distinctly, following the Ohio River, crossing the Mississippi; thence following the Missouri, on its south side, into Montana. By the increasing cold, Arctic species were driven slowly southward, generation after generation, until they occupied the whole of the United States to the Gulf, and the whole of Europe to the Mediterranean. As these species on the two continents came from a *common home* in polar regions, they were similar to one another, except in so far as some slight divergent modification may have been produced during their southward travel. When the glacial rigor declined, and the ice-sheet gradually retreated to its present position, Arctic species, following the snow-edge, went also northward, on both continents, to their present home in polar regions. But there was an alternative way of migration left open which was embraced by certain plants and insects. While on both continents most individuals went northward, some of them went upward, following the snow-edge into high mountains, and were left *stranded there*. Thus it has come to pass that the plants and insects of high mountains in temperate regions of different continents, though so widely separated and impassably isolated, are extremely similar to one another.   But, though similar, they are rarely identical.   The time has been long enough for some but not very great divergent modification.   It is impossible to conceive a more beautiful illustration of the principles we have been trying to enforce.

Thus, then, undoubtedly all the phenomena of geographical distribution of species are most rationally explained on the principle of slow evolution-changes, different in different places, and increasing with the time of isolation and its completeness.

Objection.—The only objection which can be raised against this view is the manner in which contiguous geographical faunas and floras pass into one another when they are *limited not by barriers but by temperature*.   In passing from equator to poles, over continuous land, we of course pass through many successive faunas and floras, limited wholly or mainly by temperature.   Now, if species are indeed indefinitely modifiable, then on the borders of contiguous faunas or floras, where one species disappears and another closely allied but adapted to a colder temperature takes its place, the one species (say the anti-evolutionists) ought to be gradually *transmuted* into the other, so that all the gradations may be traced.   But this is certainly not usually the fact. On the contrary, a species may indeed pass out gradually, and another come in gradually, so far as *number and vigor of individuals are concerned ;* but, in *specific character,* they may be said, usually at least, to come in suddenly, with all their characters perfect, remain unchanged throughout their whole range, and pass out suddenly at its borders. Another species takes its place, overlapping in range and coexisting on the borders of both ; this also continues unchanged, as far as it goes, and so on. The change from one fauna to another is apparently not by *transmutation* of one species *into* another by gradations, but by *substitution* of one perfect species *for* another perfect species. As a broad general statement, the condition of things is precisely such as would be the case if specific types were substantially immutable by physical conditions, but were originated in some inscrutable way (created) in the regions where we now find them, and have spread in every direction as far as physical conditions and struggle with other species would allow them— their ranges therefore interpenetrating and overlapping one another on their borders.

Two characteristic examples will make our meaning clear. There is not a more characteristic tree known than the sweet-gum, or liquidambar. This tree grows from the borders of Florida to the shores of the Great Lakes. It may indeed be most numerous and vigorous somewhere in the middle region, and may die out gradually in number and vigor of individuals on the borders of its range, but in specific character it is substantially the same throughout, easily recognizable by its dense wood, its winged bark, its five-starred leaf, its spinous burr, and its fragrant gum. Physical conditions may diminish its number and vigor, and limit its extension, but seem powerless to essentially modify its specific character. It seems to give up its life rather than change its nature.

Another striking example : The sequoias (redwood and big-tree) are entirely confined to California, and there are only two species now existing, viz., the redwood (*S. sempervirens*) of the Coast Ranges, and the big-tree (*S. gigantea*) of the Sierra Nevada. Doubtless they are most numerous and vigorous somewhere in the middle of their range, and die out gradually in number and vigor on the borders north and south, being replaced there by other genera better adapted to the physical conditions ; but in specific character they remain essentially unchanged throughout. They are everywhere the same—easily recognizable by wood, bark, leaf, and burr. Both in this case, and in the previous one of the sweet-gum, it is as if they were created perfect in their present localities, and have spread in all directions as far as physical conditions and the struggle with other competing species would allow ; but physical conditions seem powerless to change them into any other species by adaptive modification.

Answer.—We have, we believe, stated the objection fairly. The answer is, that the elements of *time* and
16

of *migrations* have not been taken into the account. In fact, this objection was conceived and formulated before the idea of geological time was fully assimilated by the human mind, and our theories of origin adjusted to it. If these species did indeed originate where we now find them, and *in the present geological epoch,* the argument might at least be entertained ; but this is not the fact. We know something of the geological history of all these species, and the history of the migrations of some of them. We know that sweet-gums were abundant and of *many species* in the United States in Tertiary times, and all have become extinct except this remnant. Whatever of modifications there were must be looked for at or about the time of its origin in Tertiary times, not now. Species, like individuals, are plastic only when young. This one has already become rigid, and all the more so as it is a remnant widely separated from other species. For competition is strongest and most effective with nearest allies. Present species are mostly isolated remnants—terminal twiglets of the tree of life. Twiglets are of course widely separated at their visible ends. Their points of union with other twiglets must be sought below.

In the case of the sequoias, we know something also of the history of their migrations. In Miocene times they were abundant, and of many species in circumpolar regions. Some twenty-four species of fossil sequoias are known, fourteen of which are Tertiary. By the cold of the Glacial epoch they were driven slowly southward, both in America and in Europe—in America as far as Southern California. After the Glacial epoch, and the return of temperate conditions, they doubtless attempted to go northward again ; but these great changes were too much for them ; they were wholly exterminated in Europe, and nearly so in America. A few were left stranded high up on the slopes of the Sierra Nevada, and on the cool, moist slopes of the Coast Ranges. The species now in California are not identical with those found in the Miocene strata of Greenland ; but the difference is only what we might expect after such extensive migrations and such long and severe struggle for life. Further, it is noteworthy that the Miocene species fall into two groups, viz., the yew-like leaved and the cypress-like leaved. These are represented to-day in California, the one by the redwood, the other by the big-tree. They are evidently direct descendants of the Miocene species, though somewhat modified.

But it will be objected that there ought to be some cases of transitional forms showing transmutation—in fact, there ought to be some cases of species now forming under our eyes. There are, we believe, examples of such cases. But intermediate forms are not likely to be maintained long, especially if migrations occur to give rise to severe conflict of forms. In that case the intermediate forms are soon eliminated, and species become distinct. This important point will be discussed more fully in the next chapter.

CHAPTER IX.

As already stated, page 40, the use of the method of experiment in the field of biology is, unfortunately, very limited. Nevertheless, it is already beginning to be used more and more in the department of physiology, and may be used also, to a limited extent, in the department of morphology. It is true that direct *scientific* experiments, for the express purpose of producing permanent modifications of form, and thus testing the theory of evolution, are of comparatively little value as yet, because the all-important element of time is wanting. The steps of evolution are so slow, and the time necessary to produce any sensible effect is usually so great, that, in comparison, man's individual lifetime is almost a vanishing quantity. But, from time immemorial, experiments have been *unconsciously* made by man on domestic animals and food-plants, which bear directly on this subject. All domestic animals and food-plants, and many ornamental flowering plants, have been subjected for ages to a process of artificial selection acting upon natural variation of off-spring. As wild species are modified, we believe, in-definitely by divergent variation and *natural* selection, so domestic species are modifiable certainly largely, perhaps indefinitely, by divergent variation and *artifi-cial* selection by man. We all know the extraordinary modifications which have thus been gradually brought about in domestic animals, such as dogs, horses, sheep, pigeons, etc.; in food-plants, as cereal grains, garden-vegetables, etc., and in ornamental plants, as roses, dahlias, pinks, etc. We can only give very briefly the principles of the process by which these extreme modi-fications are produced, referring the reader to works specially devoted to this subject for more complete ac-counts.

Let it be borne in mind, then (*a*), that inheritance is not only from the immediate parents, but from the whole line of ancestry. The inheritance from the im-mediate parents is, doubtless, usually greater than from any other *one* term of the ancestral series—the effect on the offspring of any previous generation becomes, doubtless, less and less as the distance from the off-spring increases—yet the *sum* of the ancestral inherit-ance is far greater than the immediate parental. Let it also be borne in mind (*b*) that true breeding from one form for many generations creates a fund of he-redity in that form, and thus tends to produce fixity, rigidity, or permanence in that form.

Now, the method of producing artificial breeds, sometimes consciously, sometimes unconsciously, is, briefly, as follows : Suppose it be desired to obtain a variety of an animal, say a dog, having a certain character. We start from a common type, *a* (Fig. 68). If this type

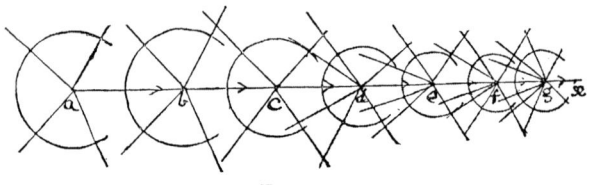

FIG. 68.

were allowed to breed naturally, the slight divergent variation of offspring represented by the radiating lines would neutralize one another by interbreeding, the individual differences would be *"pooled"* in a common stock, and the species would remain substantially constant. But if among all these slightly divergent varieties we select one, *b*, which seems in the right direction, and ruthlessly destroy all the others (indicated by crossing them out by the circular line), and breed this variety, *b*, only, we shall get again a number of divergent varieties. It may be that the larger number of these will be backward, in the direction of the original type *a*, on account of the ancestral heredity in that direction, but some will again be in the desired direction. Let all the varieties other than the desired one, but especially the backward-going or reverting ones, be again destroyed, and the one kind only selected which seems to be in the right direction, viz., *c*. As we push the form thus from generation to generation in the desired direction, especially if we attempt to hasten too much the process, the resistance to move-ment—if I may use the expression—in that direction becomes greater and greater (shown by the decreasing distances between the successive points of divergence, a, b, c, d, etc.), and the tendency to reversion becomes stronger (shown by the greater number and length of the backward-going lines), until finally it is almost impossible to push any farther. We will suppose that x is such a limit. But if, now, we breed true on the point x, destroying the reversions or backward varia-tions for many generations, we will gradually accumu-late a fund of ancestral heredity on this point which increases with every added generation, until finally the tendency to reversion becomes small. The variety *breeds true* without further interference, or with only very general superintendence. Such a permanent va-riety is called a *race*. After a race is firmly established for a sufficient length of time, and the tendency to reversion is lost, it may itself become a new point of departure for the formation of new varieties or races, in the same or other directions. Thus, during even the brief history of man, have been formed races of the different domestic animals, and useful and ornamental plants, differing so greatly from each other that, if found in the wild state, they would unhesitatingly be called different species, or even in some cases different genera.

Now, if art can vary form so greatly, and in so

short time, why may not Nature in limitless time ? If art by artificial selection, why not Nature by natural selection ? Nature is as rigid in selection and as ruthless in destruction : why may we not expect similar or even much greater results ? The process is similar in the two cases—i. e., selection among varieties in offspring, only that the selection is natural instead of artificial, and the process is so slow that there is little tendency to reversion in the latter case. Suppose, then, we have a gradually changing physical environment, or climate. Among the divergent varieties of any species in each generation, those would be preserved which are most in accordance with the new climate, and the others would perish. This is natural selection, or survival of the fittest. Add to this the effect of the change in the organic environment. All species are modified by the changing physical environment ; but these modified species again all affect one another in the competitive struggle for life, and the strongest or swiftest, or most cunning, survive (natural selection). Add to this, again, the struggle among the males for possession of the females—for reproductive opportunities—by which only the strongest and most courageous, or the most beautiful and attractive, leave progeny which inherit their peculiarities (sexual selection). Add to these, finally, *migrations*, voluntary among higher and involuntary dispersals among lower animals and plants, and the consequent mingling of faunas and floras—the migrations subjecting them to great change of environment, both physical and organic, and the mingling producing fiercer struggle for life—and we have in powerful operation many causes of modification. Add, I say, all these causes of modification together, and then make the process slow and continuous through unlimited time, and where is the limit to the degree of change? Commencing in any species, from any point of departure, there are formed first slight modifications which would be called varieties ; then these modifications, continuing in the same direction, form races ; these races by wider separation become species, and species in their turn become genera, etc. Comparing, again, to a growing tree, varieties are swelling buds ; when they grow into twigs, they are species ; when they branch again into different species, the branching stem becomes a genus, etc.

We have thus far spoken only of the various forms of one factor, viz., the Darwinian factor of selection, whether natural or artificial. We have dwelt upon this one, because the natural and the artificial processes are so similar, and the artificial is so controllable. But there are other factors in operation, in art as well as in nature. We have already spoken (p. 73) of other factors of natural change. We have shown how changing physical environment affects *function*, and function affects *form and structure*, and how these slight changes are integrated by heredity through many generations. We have also shown how *use* or *disuse* increases or diminishes the size and change the form of parts, and these changes, also, however slight, are integrated by heredity.

Now, these factors are operative also in domestication of animals and cultivation of plants. No environment is so new and peculiar as domestication and cultivation. The soil and temperature in plants, food and housing of domesticated animals, tend to change form and structure of the offspring, although in a way which it is difficult intelligently to control, and thus are prolific of varieties from which to select. In fact, they often give rise to great and unexpected modifications, called sports, which form points of departure for new varieties and races. Now, in nature, not only are all these causes and factors of change in constant operation, but they act together in a peculiarly complex way. All the members of a fauna and flora, and the physical environment of any locality, constitute together a most complex and delicately adjusted system of correlated parts. A change in one part. is propagated through the whole system ; also, a change in one factor affects all other factors. When we add to this the large amount of time, in comparison with individual human life and observation, necessary to produce visible change of form, we can easily understand why the process is still imperfectly understood, although the *fact* is certain.

But it will be asked, Are there, then, no differences between the artificially made extreme varieties equivalent, so far as difference of form is concerned, to species, and real natural species ? There are. If there were not, there would never have been any doubt about the derivative origin of natural species. But if it be asked, Are not these differences fundamental, and therefore fatal to the argument for evolution derived from this source ? we answer, we think not. We will deal frankly and fairly with these differences.

First Difference, Reversion. — The strong tendency of artificial varieties to reversion, even during the process of formation, and especially their complete reversion to the original type if the hand of man be withdrawn—i. e., if left to themselves, or become wild—is supposed to show an essential difference between such varieties, however extreme, and true species—is supposed, in fact, to prove an indestructible permanency of specific types. Nature disowns these artificial forms, and as it were brands them with bastardy. Not only so, she strives ever to destroy them. The supporting hand of man is necessary to sustain them. Left to themselves and to Nature, they quickly revert to the original type. If all the extreme varieties of dogs, from the greyhound and Newfoundland, on the one hand, to the terrier and lap-dog on the other, were turned loose on an isolated island, uninhabited by man but full of other animals, and left there to shift for themselves—and the island were visited again after a lapse of a hundred or a thousand years—it is probable that a uniform species, something like to, though perhaps not identical with, the wolf, would be found. They would have reverted to the original or nearly the original wild type from which they were produced by domestication. All or nearly all that was done by man would have been undone by Nature. This reversion is one test of species.

But the reason of this tendency to reversion is obvious : First, the time was too short, the rate of change was too rapid, in the artificial formation of these varieties. There was not time enough to accumulate a fund of heredity on each successive stage of the change. Therefore the form is unstable and the tendency to revert is strong. Compare the fleeting days and the hurrying impatience of man with the infinite time and the divine patience of Nature ! But mere instability is not the principal cause of reversion. Secondly, in the case of artificial forms in a wild state, *natural selection compels reversion.* Every species in a wild state must of course be in harmony with the environment. But artificially made forms are in harmony with the artificial environment of domestication, but not with the environment of nature. In nature the fittest survive, but artificial breeds are not fit to survive in a state of nature. They are therefore quickly destroyed in the struggle for life, or must be modified. Nature immediately begins to select the fittest, and gradually in the course of time produces one or more uniform species, similar to that from which they came, or perhaps to what they would have been by this time if left to the operation of natural causes under the conditions supposed. But natural species, if they are formed, as the derivationists suppose, by the operation of natural causes, can not revert unless the conditions revert ; for the same causes which operated to produce, still continue to operate to keep, the species. Take an example :

The form, the habits, and the instincts of the pointer have been made by a slow process of artificial selection of divergent varieties of offspring, and by training of individuals continued and its effects accumulated through many generations. But this form and these habits and instincts, so laboriously produced, would be quickly destroyed by Nature. The pointer, left to himself, must either change or become extinct, because not adapted to the wild state. Such instincts and habits would not only be of no use, but would be incompatible with success in the struggle for life. But suppose for a moment that these habits and instincts were useful to the animal in a wild state ; evidently they would be instantly seized upon by natural selection, and not only perpetuated but intensified until a very distinct species would be produced. The same is true of all other races of dogs. If the Newfoundland, the greyhound, and the pug were all turned loose in a forest, and if each of these kinds were admirably adapted to some place in the economy of Nature—for some special mode of food-getting without corresponding disabilities in other directions (as must be the case if made by natural selection)—there can be no doubt they would each survive, and their characters intensified ; intermediate forms would disappear (for reasons which we shall see presently), and we would soon have three distinct species, or perhaps we would even call them distinct genera.

Second Difference, Intermediate Forms.—Natural species are distinct—marked out with hard and fast lines—
while artificially-made races, even though in their typical
forms they differ as much or more than natural species,
shade into one another by insensible gradations.   In answer and explanation of this difference we remark : If
species or modified forms of any kind, whether natural
or artificial, are made by natural causes, and not at once
out of hand by supernatural creation, then of course
there must have been gradations in the process of making.   Now, in the artificial case, the whole process as
well as the result lies within the limits of observation,
while in the natural case only the final result.   But it
will be asked, Why are the gradations not seen also in
the final result ?   We answer, because the intermediate
forms are eliminated in the struggle for life, and not re-
produced by cross-breeding.   If artificial races always
bred true—i. e., without crossing, as natural species do—
they would probably soon be as sharply demarked.   Cross-
breeding is the great cause of the shadings between domestic races.   This brings me to the third and most important difference.

Third Difference, Cross-Fertility. — Artificially-made
races breed freely and without repugnance with one another, and the offspring of such cross-breeding is indefinitely fertile.   Natural species will not usually unite
with one another, being prevented by sexual repugnance
and other causes.   Or, if they do sexually unite, there
is either no offspring, or else the offspring is sterile,

and therefore the intermediate form dies out in the first generation ; or else the offspring is imperfectly fertile, and therefore the intermediate form is eliminated in a few generations, and the species remain distinct ; or else the offspring is more fertile with the parent stocks, and therefore revert to the parent stocks, and still the species remain distinct. Such infertile, or imperfectly fertile, offspring—the result of crossing of species—are called hybrids.

This is regarded as a most important test of true species, as contrasted with varieties or races. There are two bases on which species may be founded. Species may be based on *form*, morphological species ; or they may be based on *reproductive functions*, physiological species. By the one method a *certain amount of difference* of form, structure, and habit, *constitutes species ;* according to the other, if the two kinds breed freely with each other and the offspring is indefinitely fertile, the kinds are called varieties, but if they do not they are called species. The two tests, however, do not always accord. Every now and then we find undoubted morphological species which may be crossed and produce indefinitely fertile offspring. Yet it is certainly true that species are usually cross-sterile, while varieties, whether natural or artificial, are cross-fertile.

In explanation of this important difference, let it be observed that there are here two things which must be kept distinct in the mind, although they are, doubtless, closely allied—viz., sexual repugnance (psychological element) and cross-sterility (physiological element). The former is found, of course, only in the higher animals, where fertilization is *voluntary*. The latter is universal among all living things. This latter, therefore, is the more fundamental and essential element, and the former may be regarded as its psychical sign in the higher animals. It is of this latter, therefore— i. e., cross-sterility—that we shall speak mainly.

Suppose, then, we have growing together in the same locality many species of pines or oaks, or other ane-mophilous trees. The whole air is filled with the pollen of many species, and every germ-cell must receive many kinds of male cells, and yet there are no hybrids, but, on the contrary, the species remain distinct. So also in case of hermaphrodite animals, where the fertilization is involuntary ; many aquatic species are found together in the same locality, and the water is filled with sperm-cells of many different species. Many kinds of sperm-cells must fall on each germ-cell, and yet there are no hybrids ; the species remain distinct. In all such cases we must suppose that there is, among the different kinds of male cells, a struggle for the possession of the germ or female cell, or a sort of sexual selection by the female cell among the competing male cells, and the fittest— the most in accord ; i. e., those of the same species —prevail. This is universal. But in the higher animals, in addition to the prepotency of male cells of the same species, and comparative infertility in case of union of those of different species, sexual attraction and sexual repugnance contribute to the same result, and species are thus doubly separated. Thus sexual selection is of two kinds : selection of individuals for union (psychical), and selection of sperm-cells for fertilization (physiological). The one kind is usually the sign of the other— attraction the sign of fertility, and repugnance of sterility.

But in the domestic state it is all otherwise. Free competition between individuals or between cells is not allowed. Thus, for example, among plants, crossings may be forced and hybrids made in gardens which would never occur in Nature. The florist prevents fertilization in the same kind and compels fertilization of a different kind. If male cells of the same kind were allowed to compete, the result would be different. Doubtless the same method would succeed in many lower animals. So also in higher animals free competition and sexual selection for union are often not allowed, and therefore animals of different species, such as the horse and the ass, unite, which would not do so if they were free to select as in the wild state. These two are widely distinct species, sometimes even called genera, and therefore the offspring is infertile ; but two closely allied species, such as two species of wolf, or of the fox, in a domestic state would probably not only unite but produce indefinitely fertile offspring. In fact, it is almost certain that the dog was made by a mixture of several species of wolf, most, perhaps all, of them now extinct.*

---

* "Origin of Races of the Dog." "Annals and Magazine of Natural History," vol. xvii, p. 295. 1886.

17

On the other hand, it is not at all certain that the extreme varieties of dogs have not passed the limit of greatest attraction, and therefore of greatest cross-fertility, and that, if allowed free choice, as in Nature, they would not breed true, or tend to breed true, with their own kind, and intermediate kinds die out in the struggle for life.

Law of Cross-breeding.—Before going any further in this discussion, it is necessary to bring out another point of extreme importance in the formation of varieties, both natural and artificial—a point which I believe throws light upon the very significance of sex itself—I refer to the effect of cross-breeding.

It is a curious and most significant fact that different varieties, both natural and artificial, are, up *to a certain limit*, not only cross-fertile and cross-attractive, but even more so than individuals of the same variety. Long experience has shown that very close breeding of the same variety for a long time fixes the *kind* but *weakens* the *stock*, especially in fertility, while judicious crossing of varieties strengthens the stock, increasing its fertility, and especially producing *plasticity* or *variability*. Therefore breeders, if they wish to preserve a valuable variety, breed close ; but, if they wish to make new varieties, cross-breed. But we have already seen that species are usually cross-sterile. Therefore there must be some regular law of increase to a maximum, and again decrease to zero. It is this law that I now wish to investigate.

In the lowest animals and plants multiplication of individuals and the continuance of the kind are independent of sex, and therefore in such there may be no sex at all. The sexual elements are not yet differentiated. An individual divides itself into two; each grows to the original size and again divides into two, and so on, it may be indefinitely. In this lowest form of reproduction the individual is sacrificed to the kind, or else we may regard the kind as an extension of the individual, and reproduction as a modification of growth. But there are other sexless modes of reproduction, found in nearly all plants and many lower animals, in which the individuality is not sacrificed. The next step in the ascending scale is reproduction by *budding*. In this case a bud is formed which grows into a perfect individual, and may remain attached to the parent stalk, forming together a compound individual, as in most plants and many lower animals, such as the coral; or it may separate and assume independent life, as in some plants and many lower animals. In still other animals, as in many hydrozoa, the budding function is relegated to a special part, which thus becomes a reproductive *organ*. The next step is the placing of the budding organ, for greater safety, in an *interior cavity*. This is the case with aphids. Now, why would not this be an excellent mode of reproduction for all animals, man included? Why was sex introduced at all? There are very sufficient reasons, of many kinds, which may come up later; but the fundamental reason, in connection with evolution, is *the funding of individual differences in a common offspring, thereby giving to the offspring a tendency to divergent variation.*

Now, *non-sexual* reproduction is *absolute true breeding.* The law of like producing like is absolute. Heredity is all-powerful, and tendency to variation is *nil.* These modes of reproduction are in fact but a modification of growth and an extension of the individual. Evolution-changes in animals produced in this way only must be very slow, since the most powerful factor of evolution, viz., natural selection among divergent varieties of offspring, would be wanting. In the earliest times, therefore, before sex was yet declared, we may imagine that physical environment was the great and only factor of change. Sexual reproduction introduces the new element of variation of offspring from which Nature makes her selections ; and this element of variation is apparently the result of the union of *diverse* individuals, and the funding of these differences in a common offspring, and thus a double inheritance of individual characteristics from the parents and a multiple inheritance of the same from the ancestry. See, then, with this end in view, the pains Nature has taken to make the difference between the uniting individuals and the diversity of inheritance by the offspring as great as possible, and yet the gradual way in which she has accomplished it. As already said, the lowest form of reproduction is that by *fission.* Next comes budding in *any* part indifferently. Next comes the relegation of the budding function to a

particular part. This is the first appearance of a reproductive *organ*. Next comes the placing of this organ,
for greater safety, within. Thus far all is non-sexual
reproduction—all a modification of growth—an extension
of the individual, like the propagation of plants by cuttings and by buds. Then comes sexual reproduction in
its lowest forms.

It may be well to stop here, to show the entire difference between this and non-sexual modes. The latter, we
have seen, is only a modification of growth, an extension
of the individual. Now, sexual reproduction is the opposite of all this. Growth is a constant multiplication
of cells. One cell is ever becoming two similar cells—or,
if we call them individuals, one individual is ever becoming two similar individuals. But in sexual reproduction
we have an exactly reverse process. Reduced to its simplest terms, sexual reproduction is the fusion of *two diverse* cells, sperm-cell and the germ-cell, to form *one* cell,
the ovule—literally, a diverse twain forming one flesh.
In its higher forms it is the union of diverse *individuals*
to bring about the same result. Instead of one cell becoming two, it is two cells becoming one ; instead of one
individual becoming two in the offspring, it is two individuals becoming one in the offspring. But this great
change was not brought about at once, but only in the
most gradual manner. First, the sexual elements—sperm-
cell and germ-cell—are separated, but *in the same organ*.
Then the organs—spermary and ovary—are separated,
but in the *same individual*. This is the condition of self-

fertilizing hermaphroditism so common among plants and lower animals. Then comes cross-fertilizing hermaphroditism ; and Nature takes much pains and uses many ingenious devices to prevent self-fertilization and insure cross-fertilization. Now, for the first time, we have slight individual differences funded in a common offspring. Then, in order to absolutely forbid self-fertilization, and at the same time allow greater differences in the crossing individuals than could be attained in hermaphroditic individuals, the sex organs are separated in *different individuals,* and fertilization can only take place by *voluntary union.* Then, to insure the union of suitable individuals, and forbid the ban between unsuitable, there are introduced sexual attraction and repulsion. Then, last of all, the difference between the two sex-individuals becomes greater and greater as we go up. It is conspicuous only in vertebrates and some insects, and very conspicuous only in birds and mammals.

We see, then, as we go up the taxonomic, and undoubtedly also the phylogenic series, that there is a cross-breeding of more and more diverse individuals, a funding of more and more divergent characteristics in a common offspring. Why is this ? I answer, for the sake of *better results in the offspring.* This is abundantly shown by direct experiment. In hermaphroditic plants in which there may be either self-fertilization or else cross-fertilization with other individuals of the same species, the latter produces better results in number and vigor of offspring. But there are other advantages, more difficult to prove

but none the less certain, and of the greatest importance
in evolution : First, as already stated, complexity of inheritance, like complexity of composition in a chemical
substance, gives instability to the embryo, and thus lia-
bility to variation in the offspring ; and this in its turn
furnishes the material for selection of the fittest. Again,
it seems to me that there is a direct tendency to improve
the offspring by a sort of struggle in the embryo among
the various qualities inherited from both sides, and a
survival of the best and strongest—a sort of pre-potency
of strong qualities.

Can divergence of uniting individuals and the funding of diverse characteristics go any further ? It may.
The differences of the uniting individual may be still further increased, and the resulting offspring still further
improved by the cross-breeding of different varieties of
the same species, for we thus add varietal differences to
sexual differences in the uniting individuals. It is well
known that too close breeding, or consanguineous breeding, or breeding in and in, as it is variously called, if
continued long, has a bad effect on the offspring, weakening the stock, while judicious crossing of varieties within
certain limits of difference has a good effect, strengthening the stock and increasing its fertility. It probably
does so in two ways : one direct, by funding many diverse
qualities from both sides, and the survival in the offspring, of the strongest and best ; the other indirect, by
giving *plasticity*, instability to the embryo, and variability to the offspring, and therefore abundant material

for the operation of selection, either by man or by Nature. We said, "within certain limits of difference." If
the difference is extreme, as in extreme varieties and
races, then the effect becomes again bad, and more and
more so as the limit of specific difference is approached ;
at which limit at last Nature shuts down and forbids the
bans. Thus, then, there is in cross-breeding a regular law
of effect, increasing to a maximum and again decreasing,
which may be graphically represented by a curve (Fig.
69). In this figure the horizontal line represents the or-

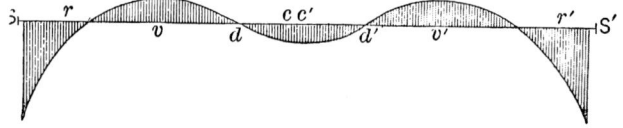

FIG. 69.

dinary level of the type ; distances on this line represent
differences, individual, varietal, or specific ; ordinates
above or below represent the effect, good or bad, of crossing. Thus s s' represent two species, and the line between
represents their specific differences; r r' represent different
races or permanent varieties ; v v' two strong varieties ;
d d' ordinary individual differences ; c c' close resembling
or consanguineous individuals. The undulating line represents the effect of crossing these various kinds. It is
seen that " in-and-in breeding," c c', produces bad effect
(negative ordinates) ; breeding of ordinary individual
differences, d d', keeps the stock at the ordinary level—in
its typical form ; crossing two strong varieties, v v', produces maximum good effect (positive ordinates) ; crossing

decided races produces again bad effects, which become infinitely bad as we approach species, s s'.*

It is generally admitted that long-continued very close breeding has a bad effect. Even in plants, Darwin has shown that cross-fertilization has better effect than self-fertilization, this last being of course the closest possible breeding. But it is probable that the principal bad effect is not on the stock but *on the process of evolution*. Very close breeding weakens the stock, ordinary breeding of individual differences maintains the stock at the ordinary level and fixes it. Cross-breeding of varieties strengthens the stock, and also (and this is its main advantage) produces plasticity in the stock, gives rise to strong divergent variations, or even sports, and thus becomes a main agent in evolution. It is probable, moreover, that the higher the function the more sensitive is it to these effects of breeding. Therefore, the effect is greater in man than in any other animal. It is true that many have doubted the bad effect of close breeding in man, and have brought forward formidable statistics to substantiate their position ; but these doubtless take no account of the most important function, the psychic, and especially the most important element in every function, so far as evolution or progress is concerned, viz., *plasticity* or capability of progressive improvement. The tendency of consanguineous breeding, or even the breeding of per-

---

* Mr. Galton (" Nature," August 26, 1886) has used a diagram similar to the above (which I first used in 1879) to illustrate the law of sexual attraction and repugnance.

sons of like character and experiences, as in an isolated community, is, if not to deteriorate the physique, at least to fix, stereotype the character, and thus to check social progress. Contrarily, the crossing of varieties of the same race seems not only to strengthen but, by the diverse inheritance, to produce plasticity of character and capacity for progress. But the difference between the primary races seems too great for crossing with advantage. Some degree of sexual repugnance which undoubtedly exists between the primary races is the psychical sign of this fact.*

If, now, we go back to what we said before taking up this subject of the effect of cross-breeding, we at once see that there is an apparent flaw in all our reasonings. If close in-and-in breeding produced better and more numerous offspring than cross-breeding between slight varieties, then, indeed, such varieties would be preserved, and increase in divergence from generation to generation until they became species. Or, in any case, if, in any way, divergence could reach the point of extreme varieties or races, or what are called sub-species, then commencing cross-sterility would complete the separation, and thus form true species. But how can the process of progressive divergence begin, when slight varieties are even more fertile by cross-breeding than by close breeding ? Is it not evident that, with every generation,

---

* This subject is more fully discussed by the author in an article entitled " Genesis of Sex," in " The Popular Science Monthly," vol. xvi, p. 167, 1879.

the slight varieties would cross-breed with one another and with the parent stock, and thus all varietal differences would be funded into a common stock, and the type would be preserved unchanged ? This, as already pointed out (p. 76), has always been the chief difficulty in the way of imagining how varieties can grow into species ; and the difficulty is *only increased* by our discussion of the law of cross-breeding. Now, just here, Dr. Romanes's most important and prolific idea comes to our help, and, as it seems to us, completely solves the difficulty.

According to Dr. Romanes, no organ is so subject to variation as the reproductive, and this in no respect so much as in degrees and kinds of fertility—we might almost say so subject to freaks of cross-sterility. Now, suppose we start with any well-defined species in a state of nature. With every generation there are many slightly divergent individual varieties, some greater and some less ; but these are all immediately swamped by crossing with one another and with the parent stock, and the species remains unchanged. But suppose among these divergent variations there arise, from time to time, some which affect the reproductive organs in such wise that the variety, though perfectly fertile with its own kind, is infertile, or imperfectly fertile, with other varieties, and especially with the parent stock. The change may be only in the *time* of flowering in plants, or season of heat in animals, or it may be actual infertility in sexual union. Right here we have the be-

ginning of a new species. The variety is sexually isolated from the parent stock by cross-sterility, and
therefore all its peculiarities, however trivial, are preserved by true breeding. Cross-breeding is necessary to
make species, but true breeding preserves them. Cross-
breeding tends ever to make varieties, but immediately
destroys them again. This constant forming and swamping, separating and again merging of varieties, like
mixing of dough, makes the whole mass (stock) more
and more plastic and subject to variety. This plasticity finally gives rise to varieties of the kind which
produces species by sexual isolation. By continued
merging the centrifugal forces continually increase, but
are continually repressed by crossing, until finally varieties break away to form species.

Now it is easy to see, from this point of view, why
artificial varieties are cross-fertile. It is because in
artificial breeding we are intent only on making varieties in form, size, color, etc., and not at all on making
any characterized by cross-sterility with the parent
stock. Cross-sterility with the parent stock, or with
other varieties, would be of no advantage, because we
control the breeding, and can breed true if we desire.
Sexual isolation is not necessary, because we can use
physical isolation. On the contrary, such cross-sterility
would be a positive disadvantage to the breeder, by
limiting the range of his experiments just where they
would be most prolific in making new varieties. Hence,
as might be expected, all domestic varieties are cross-

fertile, unless it be the extreme varieties, which may, in some instances, have passed the limit of greatest fertility.

If this idea be true, then species which have origi-nated in the same locality ought to be always cross-sterile, but species which have grown up apart, in widely separated geographical regions, ought to be sometimes cross-fertile, because they were isolated by physical not by sexual barriers. Such, Dr. Romanes thinks, is a fact. It is, however, a very important point, which ought to be carefully investigated. We say "*sometimes.*" It is probable that most geographical species also are cross-sterile ; for, although the isolation by cross-sterility of slight varieties be the main cause of the origin of species, yet a species formed by isolation of any other kind will gradually become cross-sterile with other species. Although cross-sterility be the main cause of divergence, yet divergence beyond a certain limit, however caused, will bring about cross-sterility, because the reproductive organs will partake of the general change going on in every part.

Application.—Suppose, then, a species breeding natu-rally in a wild state. Individual varieties are constantly being formed and again funded back into the common stock by cross-breeding. If the varieties thus formed be decided, the cross-breeding will strengthen the stock, and especially will preserve and increase its plasticity or tendency to variation. Finally, among the widely di-vergent varieties there is one affecting the reproductive organs of several individuals in such wise that they are infertile, or imperfectly fertile, with the parent stock, though perfectly fertile among themselves. These form a new species, which continue to increase indefinitely.

Objection answered.—This view completes the answer to an objection which is often made to evolution : " If natural species are formed by transmutation, why is it we do not find intermediate links ? Why is not organic nature made up only of individual forms, shading insensibly into each other in such wise that classification becomes a mere device to handle more conveniently complex material ? Why is it that groups, especially species, are marked out with hard and fast lines ?" We have heretofore answered this by saying that intermediate forms are eliminated. So they are, but how ? Dr. Romanes's idea of physiological selection largely answers this. It is by the funding of *ordinary* varieties into a common parental stock by crossing, and separating *specific* varieties by cross-sterility. Thus the organic field is broken up into points about which variations oscillate. As every mass of matter, when closely examined, is found to consist of aggregations about centers of *cohesive* attraction as discrete granules or crystals, and only exceptionally do we find a homogeneous vitreous structure ; even so organic forms aggregate about points of *sexual* attraction, and the whole mass consists of discrete species, and only exceptionally —i. e., in domestication—do we find insensible shadings. Now, species are the smallest aggregate of individuals, as granules are of molecules. Species are more distinctly marked out by hard and fast lines than are other taxonomic groups only because they are the *last*, going downward, that are cross-sterile—because right here is the change from cross-sterility to cross-fertility.

If this view be true, then in *the same locality* species ought to be always distinct and without shadings. If we find shadings at all, it ought to be in intermediate geographical regions, where isolation is not sexual but physical. Now, this is exactly what we find to be the fact. *Innumerable examples of such intermediate forms in intermediate geographical regions* are now known, especially among birds and reptiles, and examples have so increased in modern times, by closer study, that naturalists, especially ornithologists, have been compelled to resort to a trinomial nomenclature in order to designate these geographical sub-species.*

If any further explanation is necessary, it will probably be found in the following suggestions :

1. The number of individual varieties constantly being formed is almost infinite, but the number of places in nature is very limited. Now, among the infinite number of slight individual varieties formed with every generation, the competitive struggle will be severest between those most nearly alike, because they are competitors for the *same* place. Only one kind suc-

---

* For examples of this the reader is referred to Cope, " Bulletin of the National Museum," No. 1 ; and to Coues's " Key to North American Birds," last edition.

ceeds, viz., the fittest. Intermediate forms are, therefore, exactly those which are eliminated in the most
wholesale way. 2. Add to this the fact that, as soon
as divergence, from whatsoever cause, reaches a certain
point, sexual repugnance or cross-sterility, or both, come
in to perpetuate and increase the separation already
commenced. 3. Add to this, again, that migrations in
higher animals, and involuntary dispersals in lower animals and in plants, and the mingling together of different faunas and floras, produces a still fiercer struggle
for life, especially between natives and invaders, and
thus great numbers of forms are destroyed ; all but the
fittest are weeded out, and therefore the distinctness of
the remainder is greatly increased. Periods of great
changes of physical geography and of climate, and therefore of wide and general migrations, are also periods
of great weedings-out of unfit forms. Thus it happens
that existing faunas and floras are little else than isolated *remnants*.

To illustrate, again, by a growing tree : If all the
buds of a tree lived and grew, they would soon become
so numerous that they would together form a solid
hemispherical mass, like a coral-head, with no room
between for leaf or light or air. But ninety-nine one-
hundredths of buds die in the struggle for light and
air, and therefore the survivors are distinct growing
points, widely separated from each other. Species are
such extreme, but separated, twiglets of the tree of life.

Objection.—But it will be objected, again : The twig-

points are, indeed, separate, but the twigs themselves must meet somewhere lower down, where they began to grow. Intermediate links may be wanting *now*, but they must, of course, have existed once—i. e., in previous geological times, and therefore ought to be found fossil. In distribution in space or geographically, organic kinds may be marked off by hard-and-fast lines, but, if their derivative origin be true, in their distribution in *time* or geologically, there ought to be many examples of insensible shadings between them. In fact, if we only had all the extinct forms, the organic kingdom, taken as a whole and throughout all time, ought to consist not of species at all, but simply of individual forms, shading insensibly into each other, like the colors of the spectrum, and our classification ought to be a mere matter of convenience, having no counterpart in nature. But this is not the fact. On the contrary, the law of distribution in time is apparently similar in this respect to the law of distribution in space, already given (page 169). As in the case of *contiguous* geographical faunas, the change is apparently by *substitution* of one species *for* another, and not by *transmutation* of one species *into* another. So also in *successive* geological faunas, the change seems rather by substitution than by transmutation. In both cases species seem to come in suddenly, with all their specific characters perfect, remain substantially unchanged as long as they last, and then die out and are replaced by others. Certainly this looks much like immutability of specific forms, and

18

supernaturalism of specific origin. We have, we be-lieve, satisfactorily explained this in the case of geo-graphical distribution (page 201), but how can we ex-plain it in the case of geological distribution ?

Answer.—1. The reason for this, given by Darwin and other evolutionists, is the extremely fragmentary character of the geological record. If the existing faunas and floras are but isolated remnants, the rest having been *destroyed* by migrations and conflicts, how much more are fossil faunas and floras but fragmentary remnants, the rest having been *lost*, partly because never preserved, and partly by destruction of the record ! If from this cause existing species are widely separated, how much more ought we to expect to find fossil species distinct and widely separated !

This is undoubtedly in most cases a true and suffi-cient answer, yet we think the fragmentariness of the geological record has been overstated. While it is true that there are many and wide gaps in the record ; while it is true, also, that even where the record is continuous many forms may not have been preserved, yet there are some cases, especially in the Tertiary fresh-water de-posits, where the record is not only continuous for hun-dreds of feet in thickness, but the abundance of life was very great, and the conditions necessary for preser-vation exceptionally good. In such cases the number of fossil species found on each horizon seems to be as great as in existing faunas over equal space. The rec-ord in these cases seems to be continuous and without break, and crowded with fossil forms; and yet, although the species change greatly, and perhaps many times, in passing from the lowest to the highest strata, we do not usually, it must be acknowledged, find the gradual transitions we would naturally expect, if the change were effected by gradual transformations. The incompleteness of the record, therefore, although a true and important cause, is not the whole cause.

In further and completer answer to this greatest of all objections, we will throw out the following suggestions :

2. We must remember that considerable latitude is allowed by the anti-derivationists to *variation of species ;* so much so, indeed, that it is often difficult to draw the line between well-marked varieties and closely-allied species. Now, according to the derivationist, these strong varieties, breeding usually true, are naught else than commencing species.

3. On every side and everywhere, both in existing faunas and in fossil forms, but especially in the latter, we find innumerable examples of transitions, or intermediate forms, between all the *higher groups*, such as genera, families, orders, and classes. It is, in fact, by means of these that the great law of differentiation from generalized types has been established. It is, therefore, only between *species* that such intermediate forms are rare.

4. But even between species such intermediate forms, though rare, have been pointed out, both in existing

and in extinct faunas. But the opposition contend that, in all such cases, the previously supposed species are only varieties. We have already (page 61) spoken of the obvious fallacy involved in this position. Species are first defined as forms distinct and without intermediate links, and then we are challenged to find such links; and when, with much labor, we find them, they say the supposed species are not species, but only varieties. But there are some cases in which this subterfuge will not do. There are cases in which the transitions are between forms so extreme that they can not, by any stretch of the term, be called varieties. We will select and dwell upon but one striking example, viz., the fossil forms of the Tertiary fresh-water deposits of Steinheim.

In Würtemberg, near the little village of Steinheim, are found certain strata of sand and lime, which are evidently deposits from a small lake of Tertiary times. The deposits are extremely rich in fossil shells, especially of the different species of the genus *Planorbis*. As the deposits seem to have been continuous for ages, and the fossil shells very abundant, this seemed to be an excellent opportunity to test the theory of derivation. With this end in view, they have been made the subject of exhaustive study by Hilgendorf in 1866,* and by Hyatt in 1880.† In passing from the lowest to the highest strata

---

* "Monatsbericht d. k. Preuss. Akademie d. Wissenschaft zu Berlin," for July, 1866.

† "Genesis of Tertiary Species of Planorbis at Steinheim." A. Hyatt, Anniversary Memoir of the Boston Society of Natural History, 1880.

the species change greatly and many times, the extreme forms being so different that were it not for the intermediate forms they would be called not only different species but different genera. And yet the gradations are so insensible that the whole series is nothing less than a demonstration, in this case at least, of origin of species by derivation with modifications. The accompanying plate of successive forms (Fig. 70), which we take from Prof. Hyatt's admirable memoir, will show this better than any mere verbal explanation. It will be observed that, commencing with four slight varieties—probably sexually isolated varieties — of one species, each series shows a gradual transformation as we go upward in the strata—i. e., onward in time. Series I branches into three sub-series, in two of which the change of form is extreme. Series IV is remarkable for great increase in size as well as change in form. In the plate we give only selected stages, but in the fuller plates of the memoir, and still more in the shells themselves, the subtilest gradations are found.

This case is striking, partly because it is a very favorable one, but mainly because it has been so carefully studied. There can be no doubt that equally careful study would reveal the same transition in many other cases. Nor are such transitions confined to the lower forms of life, though they are probably more abundant there. According to Cope, the nicest gradations may be traced between some of the extinct mammalian species so abundant in the Tertiary deposits of the West—espe-

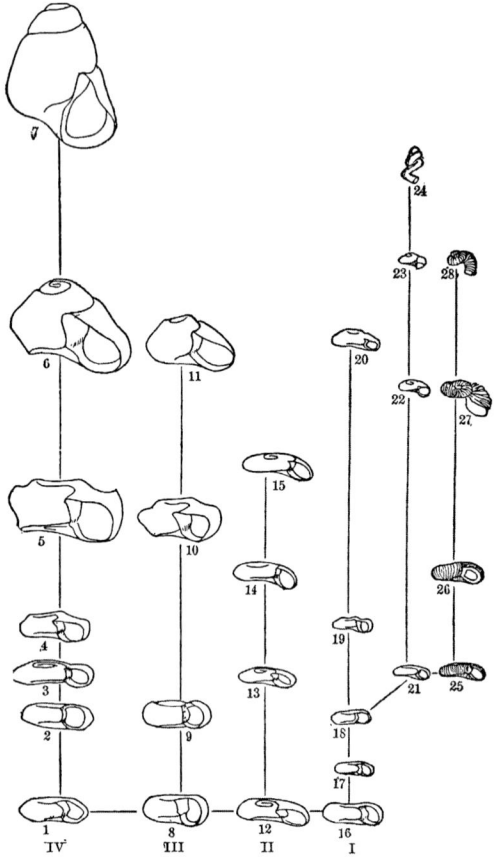

FIG. 70.—Transformations of Planorbis (after Hyatt).

Series IV. 1, Pl. levis: Undorf.   2, Pl. Steinheimensis; 3, tenuis–Steinheimensis; 4, tenuis; 5, discoideus; 6, trochiformis–discoideus; 7,
trochiformis: Steinheim.

Series III.   8, Pl. levis: Undorf.   9, Pl. oxystomus; 10, supremus; 11,
supremus var. turrita: Steinheim.

Series II.   12, Pl. levis: Undorf.   13, Pl. crescens–parvus; 14, 15, crescens: Steinheim.

Series I.   Sub-series 3.   16, Pl. levis: Undorf.   17, Pl. minutus–levis;

18, minutus; 19, 20, triquetrus: Steinheim. *Sub-series 2.* 21, Pl. minutus; 22, 23, denudatus–minutus; 24, denudatus var. distortus: Steinheim. *Sub-series 1.* 25, Pl. costatus-minutus; 26, costatus; 27, 28, costatus var — : Steinheim.
The specimens from Undorf all belong to an older Tertiary period than that at Steinheim.

cially between the species of the extinct generalized family of *Oredontidæ.** The same is probably true of the many extinct species of the horse family.

It is interesting to observe that the details of the process of change in the forms of *Planorbis* are in accord with Dr. Romanes's views. The change does not seem to have been uniform but somewhat paroxysmal. The forms seem to remain stable for a long time, and then a few break into several different forms, while the more rigid die out. It is as if cross-breeding had kept the type true, but at the same time increased its tendency to variation, until finally one or more varieties became sexually isolated and thus formed new species.

5. But still the question remains : Why are transitional forms *rare* in all cases, especially between species— so rare that they are eagerly sought and highly prized ? I believe that the true reason of this is that *the steps of evolution are not always uniform.*

Nearly all evolutionists have assumed and even insisted on uniformity, as the opposite of catastrophism and of supernaturalism, and therefore as essential to the idea of evolution. They say that the constancy of the

* In a letter to the author, dated February 13, 1887, Prof. Cope says: "Such transitions of species are clearly indicated in the *Oreodontidæ,* where such different forms as *O. gracilis* and *O. Culbertsoni* are connected by intergradations."

action of the forces of change necessitates the uniformity of the rate of change. But, in fact, this is not always nor even usually true. Causes or forces are constant, but phenomena everywhere and in every department of Nature are *paroxysmal*. The forces producing storms and lightning, and volcanoes and earthquakes, are or may be constant; yet the phenomena are in the highest degree paroxysmal. Wherever in nature we have a constant force and a strong resistance, we find more or less paroxysmal action. For this reason the wind blows in puffs, the friction of wind on water produces waves, water running in small pipes issues in pulses. The reason is obvious, as may be seen by the following examples: Suppose lifting forces within the earth are resisted by crust-rigidity. The forces accumulate uniformly until the resistance gives way, and suddenly we have an earthquake. Water running with great resistance in small pipes is checked, but soon accumulates additional force, which overcomes the resistance, only to be again checked, and so on, and therefore runs in pulses. Now, the course of evolution of the whole earth may be likened to such a current; there are forces of movement and forces of resistance—progressive forces and conservative forces. The progressive force is accumulative, the resisting force is constant. Thus, in all evolution or history, whether of the earth or of society, there are periods of comparative quiet, during which the forces of change are gathering strength, and periods of revolution or rapid change, during which these forces show themselves in conspicuous effects.

Now, that there have been such periods of rapid revolutionary change in the history of the earth, there can be no doubt. The history of the earth is marked by periods of comparative quiet, during which life was exceptionally abundant and prosperous, and change of organic forms slow and uniform—separated by periods of disturbance, revolution, rapid changes of physical geography and climate, and consequently of comparatively rapid and sweeping changes in organic forms. These form the division-lines between great eras of the earth's history, and are always marked by extensive unconformity of the strata, showing the changes of physical geography above spoken of, and by apparently sudden and sweeping change in life-forms, showing the great changes of climate and other physical conditions. Unfortunately, in all cases of unconformity of strata, there is, of course, a break in the continuity of the record ; and when the unconformity is very general a portion of the record may be irrecoverably lost. The consequence is, that there is an apparent break also in the continuity of life-forms. It looks, at first sight, like wholesale extermination of old and recreation of new forms. But undoubtedly the break in the continuity of life is apparent only, as is shown by the loss in the record. If we could recover the whole record, as indeed we sometimes do, we should find in all cases that there is no break in the continuity of evolution, but only more rapid rate of change at these times. But to this cause of rapid rate of progress—i. e., change of physical environment—we must add change of organic environment induced by the physical. We have already seen (p. 179) that extensive changes in physical geography and climate are always accompanied by wide migrations and dispersals of species, the mingling of faunas and floras, and the severer struggle for life, and the sweeping weeding-out of all but the fittest, and the change of these latter, making them still fitter. These two causes of rapid change, viz., change of climate and migrations, together with the loss of record, we believe completely account for those sweeping changes, not only of species but even of genera, families, and orders which characterize the passage from one great era to another.

But this does not yet explain the apparent discontinuity between *consecutive* species in the same locality in continuous, conformable strata, or the rarity of transitional forms when one species takes the place of another in an apparently continuous record. In such continuous deposits the successive faunas do indeed gradate insensibly into one another, but apparently as in contiguous geographical regions (p. 200) by substitution, not by transmutation. How shall we explain this ?

On this point I throw out some suggestions : 1. In the modification of species, too, as well as in other progressive changes, we may imagine two forces operating, one progressive, the other conservative—the one external, the other internal. The external progressive force consists of all the factors of change already mentioned, the internal conservative is the law of heredity, of like producing like. A changing environment tends continually and increasingly to change of organisms, but change is resisted by heredity, which tends to adhere, within narrow limits, to the same form. But since the external-force or tendency to change increases constantly—since the discord between the environment and the organism becomes ever greater, there must come a time when either the species is destroyed, or else the resistance of heredity gives way, and rapid change takes place. The alternative is presented to the species to transform or perish; and in one or perhaps in two or three generations we have an amount of change which, under other circumstances, might take a hundred generations to accomplish. These rapid changes are in fact exactly what in artificial varieties we call sports. We do not know all the conditions which determine sports in domestication, and still less what determines large and widely-divergent variations, and therefore rapid origin of many divergent species, in geological history. But one thing seems probable, viz., that, when a species begins to change, it continues to change easily and in many directions. When resistance gives way it takes some time, many generations, for heredity to gather force again. Hence, young species are plastic, fluent, because heredity, on any one point, has not yet accumulated. But as soon as a stable form is again reached, then, by accumulating a fund of heredity, the form tends to become more and more rigid, until often it becomes too rigid to yield to modifying influences, and therefore becomes extinct. By far the greater number of species do thus become extinct and leave no progeny, while the few more plastic forms are modified in several directions, and the number of forms may, after a little time, be undiminished or even increased.

2. As to the *cause* of rapid changes of form during revolutionary or critical periods in the earth's history, Brooks has introduced an idea which is very suggestive, and deserves serious attention. We have above spoken of the progressive element as external. Brooks regards both elements as internal, and represented by the two 'sexes. The male represents the progressive, the female the conservative element. The one tends to divergent variation, the other to fixity of type by heredity. I think we will all admit that, as a general rule, in man (and probably all the higher animals) the male is more highly differentiated into many divergent forms—the female is more like the type-form of the species. In man, the male is certainly more diversified in form, in expression, and in character. If they have the keenest ear for musical pitch, they are also most often music-deaf; if they have the sharpest perception of color, they are also most often color-blind ; if among them we find the brightest intellects, we also find the dullest and most stupid ; if there are among them more geniuses, so, also, there are more cranks. The same is also, prob- ably, true of other animals, in proportion to their grade of organization. The operation of these two equally necessary elements is well shown in every advancing society. The initiative of every movement, in all direc- tions, good or bad, is determined by the male ; the con-

servation of whatever balance of good there may be, seems to be mainly by the female. The male tries all things, the female holds fast that which is good. By the one society gains a little in each generation ; by the other the gain is conserved and made a new point of departure. The one is ever building hastily a scaffolding and platform ; the other ever consolidating into a permanent structure. Now, according to Brooks, what is true in the plane of social progress is true also in the lower plane of organic evolution. In sexual union, and in the resulting offspring, the sperm-cell is the element which tends to divergent variation, and the germ-cell to fixity of type, through heredity. In artificial breeding, then, we ought to make new varieties by proper use of the sire ; we ought to preserve them true by proper management of the dam.

But, again, it is believed that in many lower animals, especially insects, the high-feeding of the mother, and consequent good condition of the ovum, tends to the production of female offspring. It seems almost certain that, in butterflies, the sex is not yet declared in the caterpillar stage. According to the careful experiments of Mrs. Treat,* if the caterpillars be well fed, they become female butterflies ; but, if poorly fed, they make males. One purpose of this provision of Nature is, doubtless, to provide for the greater draught on the vitality of the female in reproduction.

---

* "American Naturalist," 1873 ; "Popular Science Monthly," June, 1873.

Now for the application. In good times in the history of a species, when everything is prosperous, external conditions are favorable, and food is abundant, females are in excess, and individuals are greatly multiplied. Under these conditions, evolution would be slow and uniform. But in *bad* times in the history of a species, when external conditions were unfavorable, not only would there be excess of males, but these, through the influence of the changing environment, as well as through the dominance of the male element, would be more than usually varied in character. Among the strongly divergent varieties thus formed, the fittest—i. e., those most in accord with the changing environment—would survive and leave offspring partaking of their character. We have already repeatedly said that the severer pressure of a rapidly-changing environment determines correspondingly rapid changes in organic forms. It may do so in many ways; but, according to Brooks, one of the most important ways is by determining an excess of the male element.

In brief, then, the causes of rarity of transitional forms among fossils are—1. The change being, for the reasons given, comparatively rapid, the *number of generations* between consecutive species are few, perhaps only one. 2. Times of rapid change are also times of unfavorable conditions, and therefore the number of *individuals* in each generation is small, and all the smaller—in Brooks's view—because of the fewness of females. When we remember that fossils are but a small fraction of the actual faunas and floras of the time, surely these two causes go far toward explaining the rarity of links between *species*. 3. Add to these the existence of periods of wide-spread changes in physical geography and climate, and consequent wide migrations and dispersals of species, and we sufficiently account for those sweeping changes in species, genera, families, and orders, which mark the limits of the great eras, and which are made still more abrupt, and apparently supernatural, by the loss of record at these times.*

Objection.—There is still one more objection which will be made. We have drawings of plants, animals, and men, by Egyptian artists, who lived at least three thousand years ago, and the species of the one and the races of the other are still the same. Still better, we have among the wrappings of Egyptian mummies the very plants themselves, leaves and flowers perfectly preserved, and even colors almost perfect. Yet the species are exactly the same as grow in Egypt to-day. If species are made by gradual transmutation, surely there ought to have been some change in three thousand years.

Answer.—It may be well to note that this apparent permanence is true of *races* of men as well as of *species* of animals and plants. But the very men who

---

* For a fuller development of this subject the reader is referred to an article by the author, entitled " Critical Periods in the History of, and their Relation to, Evolution " (" American Journal of Science," vol. xiv, p. 99, 1877).

insist on permanence of species are equally insistent on the variability of varieties and races. The objection, therefore, proves too much. We shall not insist on this, however, because as derivationists we regard races as naught else than commencing species, and therefore subject to the same laws. We are not striving for triumph in debate, but only for truth. The true answer will, we believe, be found among the following suggestions :

1. Three thousand years seems a long time in human history, but in geological history it is but a day. This, the usual answer, is no doubt a true one, but hardly, we think, sufficient. When we remember the enormous change which has taken place in faunas and floras since the end of the Tertiary, if change still continues at the same rate, surely it ought to be distinctly perceptible in three thousand years.

2. But we must remember that such changes are usually more or less paroxysmal; not, indeed, so sudden as to break the continuity of life, but far more rapid at some times than at others. The last critical or revolutionary period of rapid change was the Glacial epoch. Since that time—i. e., during the human period—a new equilibrium has been established, and the changes in organic forms have been very slow.

3. Remember, again, that in evolution *all* species do not change. On the contrary, most become rigid, and either remain unchanged, almost indefinitely, or else die out and leave no descendants. Only the more

plastic forms change into other species, but usually into several other species, and thus the number of forms may be undiminished, even though the larger number of old forms leave no descendants. It is true, therefore, of this as well as of other epochs, that the greater number of species are permanent.

4. It is not impossible—indeed, it is in exact accordance with the laws of evolution—that organic forms are more permanent now than ever before. Evolution is a growth; the forces of growth must exhaust themselves. Evolution proceeds by constant differentiation and specialization, but extreme specialization always arrests evolution. In ontogenic evolution, for example, cell-structure becomes more and more specialized, but also thereby more and more rigid, and, when specialization is complete, evolution stops, and cell-forms are permanent. It is this which limits the cycle of every evolution. So is it precisely with evolution of the organic kingdom, except that the cycle is much longer. Here, also, every step is by specialization, and yet specialization fixes the form, and finally arrests the advance on that line. Thus, throughout the whole geological history of the earth, the larger number of forms, by specialization, become rigid and perish, while the fewer, more generalized, and more plastic forms take up the march and carry it forward a step, only to be themselves specialized and fixed. If we compare, again, to a tree : each twig finishes its growth, flowers, fruits, and dies ; other buds take up the growth and carry it for-

19

ward. By specialization the highest condition of a certain form of life is attained, but other possibilities are shut off. Extreme specialization is the flowering and fruiting—the end and completion of twig-life. Now, obviously, this specialization and respecialization can not go on forever. When it is complete in every direction it must cease, and forms become permanent, or else perish. When it flowers it must die.

Now, is not the advent of man in many ways a sign of the completeness of organic evolution ? Certain it is that with man there begins an entirely new form of evolution. Certain it is that with man evolution is transferred from the organic to the social plane, from the material to the psychical. Certain it is that the forces, the conditions and results of this evolution, are wholly different from those of the other. In organic evolution the organism must conform to the environment ; in human evolution the environment is made to conform to the wants of the organism. The one is unconscious and involuntary, passive under the dominating laws of Nature ; the other is conscious, voluntary progress toward an ideal, *by the use,* among other means, of the laws of Nature. The one is by change of external form—i. e., change of species—the other by change of brain-structure. Now, does not the commencing of the cycle of this new evolution imply the closing of that of the old ? The two may overlap somewhat now, but it is evident that, when the cycle of human evolution culminates, when highly civilized man shall have taken possession of the whole earth, the whole organic kingdom must be readjusted to his wants. All organic forms must be either domesticated or destroyed. Organic forms will no longer be modified by natural but wholly by artificial selection.

There are many other supposed objections which have been urged, but these are mostly not objections to evolution, but only to some *special theory* of evolution—Darwinian, Spencerian, Lamarckian, or other.

Origin of Beauty.—For example, it has been urged that natural selection can only account for *useful* structures ; but *beauty* is as universal and as conspicuous in nature as *use*. In many cases Darwin has shown that beauty is useful, and in such cases it is, of course, seized upon by selection and intensified. Thus, the gorgeous coloring of birds and insects is largely due to sexual selection. Beauty is attractive, and therefore the most beautiful prevail in securing reproductive opportunities. This character is, therefore, perpetuated in the offspring, and intensified from generation to generation. But, of course, this can apply only to higher animals, in which the sexes are separate and sexual union voluntary. It can not apply to self-fertilizing hermaphrodites ; and yet in these, also, we often find the most gorgeous coloring. Again, Darwin has very ingeniously and successfully explained the case of the beauty and fragrance of flowers of hermaphroditic plants by another principle, viz., that of *insect-selection*. Insects are attracted by the most showy and fragrant flowers, and thus become the means of carrying pollen from flower to flower, insuring fertilization, and especially cross-fertilization. The most beautiful and fragrant flowers are most certain to be fertilized, and thus beauty and fragrance become useful to the plant, and therefore are selected and intensified.

These and many other cases of beauty may doubtless be explained by showing that it is useful; but beauty which is without any use can not be explained by natural selection. Now, as already said, the most gorgeous beauty is lavishly distributed even among the lowest animals, such as marine shells and polyps, where no such explanation is possible. The process by which such beauty is originated and intensified is wholly unknown to us.

Incipient Organs.—Again, Mivart has drawn attention to another difficulty in the way of natural selection as an explanation even of useful organs. Darwin does not, of course, attempt to account for the *origin* of varieties. As we have already seen, he assumes divergent variation of offspring as the necessary material on which natural selection operates. He who shall explain the origin of varieties will have made another great step in completing the theory of evolution. But not only does not natural selection explain the *origin* of varieties, but neither can it explain the *first steps* of advance toward usefulness. An organ must be already useful before natural selection can take hold of it to improve it. It can not make it useful, but only *more useful*. For example, if fins commenced as buds from the trunk, it is difficult to see how they could be of any use, and therefore how they could be improved by natural selection until they were of considerable size, and especially until muscles were developed to move them. Until that time they would seem to be a hindrance to be removed by natural selection, instead of a use to be preserved and improved. It would seem that many organs must have passed through this *incipient stage*, in which their use was prospective.

Much that is very interesting might be said on these and similar points of difficulty, but all this lies entirely aside from the scope of this work. As already said, these are not objections to evolution or derivation, but only to *Darwinism*, or any other special theory, as a *sufficient explanation* of the process of evolution. They only show that we do not yet fully understand this process; that there are still other and perhaps greater factors of evolution than is yet dreamed of in our philosophy.

In the foregoing chapters on special evidences, and especially in the last two, the reader will observe many points of doubt, discussion, and difference of opinion. Let it not be concluded on that account that the *law of evolution* is still in the region of uncertainty. It can not be too strongly insisted on that the fact of evolution as a universal law must be kept distinct from the causes, the factors, the conditions, the processes, of evolution. The former is certain, the latter are still imperfectly understood.

PART III.

THE RELATION OF EVOLUTION TO RELIGIOUS THOUGHT.

CHAPTER I.

FROM what has preceded, the reader will perceive that we regard the law of evolution as thoroughly established. In its most general sense, i. e., as a law of continuity, it is a necessary condition of rational thought. In this sense it is naught else than the universal law of necessary causation applied to forms instead of phenomena. It is not only as certain as—it is far more certain than—the law of gravitation, for it is not a contingent, but a necessary truth like the axioms of geometry. It is only necessary to conceive it clearly, to accept it unhesitatingly. The consensus of scientific and philosophical opinion is already well-nigh, if not wholly, complete. If there are still lingering cases of dissent among thinking men, it is only because such do not yet conceive it clearly—they confound it with some special form of explanation of evolution which they, perhaps justly, think not yet fully established. We have sometimes in the preceding pages used the words evolutionist or derivationist; they ought not to be used any longer. The day is past when evolution might be regarded as a school of thought. We might as well talk of gravitationist as of evolutionist.

If, then, evolution as a law be certain, if, moreover, it is a law affecting not only one part of Nature—the organic kingdom—and one department of science—biology—but the whole realm of Nature and every department of science, yea, every department of thought, changing our whole view of Nature and modifying our whole philosophy, the question presses upon us, "What will be its effect on religious belief, and therefore on moral conduct?" This is a question of gravest import. To answer it, however imperfectly, is the chief object of this work. Except for this, it would probably never have been undertaken. All that goes before is subsidiary to this.

But I will doubtless be met at the very threshold by an objection from the scientific side. Some will say— because it is the fashion now to say—that as simple, honest truth-seekers, we have nothing to do with its effect on religion and on life. They say we must follow Truth wherever she leads, utterly regardless of what may seem to us moral consequences. This I believe is a grave mistake, the result of a reaction, and on the whole a wholesome and noble reaction, against the far more common mistake of sacrificing truth to a supposed good. But the reaction, as in most other cases, has gone much too far. There is a true *philosophic* ground of justification for the reluctance with which even honest truthseekers accept a doctrine which seems harmful to society. Effect on life is, and ought to be, an important element in *our estimate of the truth of any doctrine.* It is necessary for me to show this, in order to justify this part of my work.

Relation of the True and the Good.—There is a necessary and indissoluble connection between truth and usefulness. We all at once admit this connection in one direction. We all admit that a truth must eventually have its useful application. It may not be *now,* nor in ten years, nor in a century, nor even in a millennium, but some time in the future it will vindicate its usefulness. No truth is trivial or useless in its relation to human life, for man is a part of Nature, and his life must be in accordance with the laws of Nature. Every one admits this, but not every one admits the converse proposition, viz., that whatever doctrine or belief, in the long run and throughout the history of human advancement, has tended to the betterment of our race, must have in it an element of truth by virtue of which it has been useful ; for man's good can not be in conflict with the laws of Nature. Also, whatever in the long run and in the final outcome tends to the bad in human conduct, ought to be received, even by the honest truth-seeker, with distrust as containing essential error. The reason of this will now be further explained.

Relation of Philosophy to Life.—There are three primary divisions of our psychical nature, viz., sensuous, intellectual, and volitional or moral. There are three corresponding primary processes necessary ,to make a complete rational and satisfactory philosophy : (1) There is first the *instreaming* of the external world through the senses, as impressions. These we call facts or phenomena. (2) The elaboration of these facts within, by the *intellect*, into a compact, consistent structure. This we call knowledge. (3) The outgoing of this knowledge by the *will* into the world as right or wise conduct. Now these three are all equally necessary. All these three portions of our complex nature are equally urgent to be satisfied.* But, unfortunately, scientific workers are too apt to think only 1 and 2 necessary—that true facts elaborated into consistent theory are all we need care for. Theologians and metaphysicians, on the other hand, seem to think only 2 and 3 necessary. They elaborate a theory consistent in all its parts, exquisitely woven in beautiful and delicate pattern, and apparently satisfactory in its application to the right conduct of life, but are less careful to inquire whether it is in harmony with facts derived from the senses. But, we repeat, all three are equally necessary. The first gathers the materials, the second constructs the edifice, the third, by *use*, by practical application, *tests* whether it be a fit building to live in, whether it is constructed on sound architectural principles. The tendency of the olden time was to neglect the first, the tendency of the present time is to neglect the third. But we repeat with stronger emphasis that this third element is equally necessary. All admit that suc-

---

* " Reflex Action and Theism," William James, " Unitarian Review " for November, 1881.

cessful application in art is the surest test of the truth of science. Now, social conduct is the art corresponding to our philosophy of life, and therefore is the sure test of its truth. It follows, therefore, that unless all these three primary divisions of our nature are satisfied by any doctrine, there must result an ineradicable confusion and discord in our psychical nature, and cordial acceptance is not only impossible but irrational. We insist upon this the more because it has become the fashion in these latter days of dominance of science, to say that to inquire into effects on society is inconsistent with the scientific spirit, and unworthy of the honest truth-seeker. But, observe, I am speaking of effects on society only as a *test of truth.* I would not swerve a hair's breadth from absolute devotion to truth. It is necessary, indeed, to inquire into effects on society, but we must inquire only in the patient spirit characteristic of the truth-seeker. Whatever is really true will surely vindicate itself by its beneficence, if we will only wait patiently for final results. Evolution is no exception to this universal truth. It will surely vindicate its beneficence, but we must wait yet a little while—not very long.

So much it was necessary to say in justification of the inquiry which constitutes this third part of our work. But, after this justification, the question returns with additional emphasis, " What will be the effect of the universal acceptance of the law of evolution on religious thought, and through this on the right conduct of life ?"

There can be no doubt that evolution, as a law affect-ing all science and every department of Nature, must fundamentally affect the whole realm of thought, and pro-foundly modify our traditional views of Nature, of God, and of man. There can be no doubt that we are now on the eve of a great revolution. But, as in all great revo-lutions, so in this, the first fears as to its effects are greatly exaggerated. To many, both friends and foes of Christianity, evolution seems to sweep away the whole foundation, not only of Christianity, but of all religion and morals, by demonstrating a universal materialism. Many are ready to cry out in anguish, " Ye have taken away our gods, what have we more ? Ye have destroyed our dearest hopes and noblest aspirations, what more is left worth living for ? " But I think all who are at all familiar with the history of the so-called conflict between religion and science will admit this is not the first time this cry has been raised against science. They have heard this danger-cry so often that they begin to regard it as little more than a wolf-cry—scientific wolf in the religious fold. It may not be amiss, then, to stop a mo-ment to trace rapidly the main points of this conflict—to discuss the various forms of this scientific wolf.

First, then, it came in the form of the *heliocentric theory of the planetary system*. We once thought the earth the center of the universe, and so firm that it can not be moved. But science shows that it moves about the sun, and spins unceasingly on its axis. Every one has heard of the terror of the sheep produced by this discovery, and the nearly tragic results to the bold scientist. But now we look back with wonder that there should have been any trouble at all. Would any Christian now consent to give up the grand conceptions of Nature and of God thus opened to the human mind—the idea of infinite space full of worlds, of which our earth is one, moving in silent harmony as in a mystic dance? Verily, this wolf has proved itself a harmless, nay, a very noble beast, and lies down in peace with the lambs.

Next, it came in the shape of the *law of gravitation,* as sustentation of the cosmos by law and resident forces. The effect of this on religious thought was even more profound, though less visible on the surface, because only perceived by the most intelligent. It seemed at that time to remove God from the course of Nature. This was the real ground of the skepticism of the last century, and also the real motive of Voltaire's ardent advocacy of Newton's views before these were generally accepted in France. But now, who would give up this grand idea— this conception of law pervading infinite space—the same law which controls the falling of a stone guiding also the planetary orbs in their fiery courses? This is indeed the divine spheral music, inaudible but to the ear of science, accompanying the celestial dance.

Next, it came in the form of the *antiquity of the earth* and of the cosmos. The earth which we had fondly thought made specially for us about six thousand years ago; sun, moon, and stars, which we had vainly imagined shone only for our behoof—these, science tells us, existed and each performed its due course inconceivable ages before there was a man to till the ground or contemplate the heavens. Some of my readers may still remember the horror, the angry dispute which followed the promulgation of these facts. But now, who would consent to give up the noble conception of infinite time thus opened to the human mind and become forever the heritage of man ?

Next, it came in the form of the *antiquity of man.* It is probable, nay, certain, that man has inhabited the earth far longer than we had previously supposed we had warrant for believing. The controversy on this question and the dread of its result has indeed not yet entirely subsided. Some timid people still look askance at this wolf, but I think all intelligent people accept it and find it harmless.

Next, and last, it comes now in the form of *evolution* —of the origin of all things, even of organic forms, by *derivation*—of *creation by law.* We are even now in the midst of the terror created by this doctrine. But what is evolution but law throughout infinite time ? The same law which now controls the development of an egg has presided over the creation of worlds. Infinite space and the universal law of gravitation ; infinite time and the universal law of evolution. These two are the grandest ideas in the realm of thought. The one is universal sustentation, the other universal creation, by law. There is one law and one energy pervading all space and stretching through all time. Our religious philosophy has long ago accepted the one, but has not yet had time to re-adjust itself completely to the other. A few more years, and Christians will not only accept, but love and cherish this also for the noble conceptions it gives of Nature and of God.

But some will exclaim, "Noble conceptions of God, say you! Why, it utterly obliterates the idea of God from the mind. All other conflicts were for outworks—this strikes at the citadel. All others required only re-adjustment of claims, rectification of boundaries betwixt science and religion—this requires nothing less than un-conditional surrender. Evolution is absolute material-ism, and materialism is incompatible with belief in God, and therefore with religion of any kind whatsoever!" Before proceeding any further, it becomes necessary to remove this difficulty out of the way.

CHAPTER II.

IT is seen in the sketch given in the previous chapter that, after every struggle between theology and science, there has been a readjustment of some beliefs, a giving up of some notions which really had nothing to do with religion in a proper sense, but which had become so *associated* with religious belief as be to confounded with the latter—a giving up of some line of defense which ought never to have been held because not within the rightful domain of theology at all. Until the present the whole difficulty has been the result of misconception, and Christianity has emerged from every struggle only strengthened and purified, by casting off an obstructing shell which hindered its growth. But the present struggle seems to many an entirely different and far more serious matter. To many it seems no longer a struggle of theology, but of essential religion itself—a deadly life-and-death struggle between religion and materialism. To many, both skeptics and Christians, evolution seems to be synonymous with blank materialism, and therefore cuts up by the roots every form of religion by denying the existence of God and the fact of immortality. That the enemies of religion, if there be any such, should assume and insist on this identity, and thus carry over the whole accumulated evidence of evolution as a demonstration of materialism, although wholly unwarranted, is not so surprising; but what shall we say of the incredible folly of her friends in admitting the same identity!

A little reflection will explain this. There can be no doubt that there is at present a strong and to many an overwhelming tendency toward materialism. The amazing achievements of modern science; the absorption of intellectual energy in the investigation of external nature and the laws of matter have created a current in that direction so strong that of those who feel its influence—of those who do not stay at home, shut up in their creeds, but walk abroad in the light of modern thought—it sweeps away and bears on its bosom all but the strongest and most reflective minds. Materialism has thus become a fashion of thought; and, like all fashions, must be guarded against. This tendency has been created and is now guided by science. Just at this time it is strongest in the department of biology, and especially is evolution its stronghold. This theory is supposed by many to be simply demonstrative of materialism. Once it was the theory of gravitation which seemed demonstrative of materialism. The sustentation of the universe by law seemed to imply that Nature operates itself and needs no God. That time is passed. Now it is evolution and creation by law. This will also pass. The theory seems to many the most materialistic of all scientific doctrine only because it is the *last* which is claimed by materialism, and the absurdity of the claim is not yet made clear to many.

The truth is, there is no such necessary connection between evolution and materialism as is imagined by some. There is no difference in this respect between evolution and any other law of Nature. In evolution, it is true, the last barrier is broken down, and the whole domain of nature is now subject to law; but it is only *the last;* the march of science has been in the same direction all the time. In a word, evolution is not only not identical with materialism, but, to the deep thinker, it has not added a feather's weight to its probability or reasonableness. Evolution is one thing and materialism quite another. The one is an established law of nature, the other an unwarranted and hasty inference from that law. Let no one imagine, as he is conducted by the materialistic scientist in the paths of evolution from the inorganic to the organic, from the organic to the animate, from the animate to the rational and moral, until he lands, as it seems to him, logically and inevitably, in universal materialism—let no such one imagine that he has walked all the way in the domain of science. He has stepped across the boundary into the domain of philosophy. But, on account of the strong tendency to materialism and the skillful guidance of his leaders, there seems to be no such boundary; he does not dis-

tinguish between the inductions of science and the in-
ferences of a shallow philosophy ; the whole is accredited
to science, and the final conclusion seems to carry with
it all the certainty which belongs to scientific results.
The fact that these materialistic conclusions are reached
by some of the foremost scientists of the present day
adds nothing to their probability. In a question of science, viz., the law of evolution, their authority is de-
servedly high, but in a question of philosophy, viz.,
materialism, it is far otherwise. If the pure scientists
smile when theological philosophers, unacquainted with
the methods of science, undertake to dogmatize on the
subject of evolution, they must pardon the philosophers
if they also smile when the pure scientists imagine that
they can at once solve questions in philosophy which
have agitated the human mind from the earliest times.
I am anxious to show the absurdity of this materialistic
conclusion, but I shall try to do so, not by any labored
argument, but by a few simple illustrations.

1. It is curious to observe how, when the question
is concerning a work of Nature, we no sooner find out
how a thing is made than we immediately exclaim : " It
is not made at all, it became so of itself ! " So long as
we knew not how worlds were made, we of course concluded they must have been created, but so soon as science showed *how* it was probably done, immediately we
say we were mistaken—they were not made at all. So
also, so long as we could not imagine how new organic
forms originated, we were willing to believe they were

created, but, so soon as we find that they originated by
evolution, many at once say, "We were mistaken ; no
creator is necessary at all." Is this so when the question
is concerning a work of man ? Yes, of one kind—viz.,
the work of the magician. Here, indeed, we believe in
him, and are delighted with his work, until we know
how it is done, and then all our faith and wonder cease.
But in any honest work it is not so ; but, on the contrary, when we understand how it is done, stupid wonder
is changed into intellectual delight. Does it not seem,
then, that to most people God is a mere wonder-worker,
a chief magician. But the mission of science is to show
us how things are done. Is it any wonder, then, that to
such persons science is constantly destroying their superstitious illusions ? But if God is an honest worker, according to reason—i. e., according to law — ought not
science rather to change gaping wonder into intelligent
delight—superstition into rational worship ?

2. Again, it is curious to observe how an *old truth*,
if it come only in a *new form*, often strikes us as something unheard of, and even as paradoxical and almost
impossible. A little over thirty years ago a little philosophical toy, the gyroscope, was introduced and became
very common. At first sight, it seems to violate all mechanical laws, and set at naught the law of gravitation
itself. A heavy brass wheel, four to five inches in diame-
ter, at the end of a horizontal axle, six or eight inches
long, is set rotating rapidly, and then the free end of the
axis is supported by a string or otherwise. The wheel re-

mains suspended in the air while slowly gyrating. What mysterious force sustains the wheel when its only point of support is at the end of the axle, six or eight inches away? Scientific and popular literature were flooded with explanations of this seeming paradox. And yet it was nothing new. The boy's top, that spins and leans and will not fall, although solicited by gravity, so long as it spins, which we have seen all our lives without special wonder, is precisely the same thing.

Now, evolution is no new thing, but an old familiar truth ; but, coming now in a new and questionable shape, lo, how it startles us out of our propriety ! Origin of forms by evolution is going on everywhere about us, both in the inorganic and the organic world. In its more familiar forms, it had never occurred to most of us that it was a scientific refutation of the existence of God, that it was a demonstration of materialism. But now it is pushed one step farther in the direction it has always been going —it is made to include also the origin of species—only a little change in its form, and lo, how we start ! To the deep thinker, now and always, there is and has been the alternative—materialism or theism. God operates Nature or Nature operates itself ; but evolution puts no new phase on this old question. For example, the origin of the individual by evolution. Everybody knows that every one of us individually became what we now are by a slow process of evolution from a microscopic spherule of protoplasm, and yet this did not interfere with the idea of God as our individual maker. Why, then, should the discovery that the species (or first individuals of each kind) originated by evolution destroy our belief in God as the creator of species ?

3. It is curious and very interesting to observe the manner in which vexed questions are always finally settled, if settled at all. All vexed questions—i. e., questions which have tasked the powers of the greatest minds age after age—are such only because there is a real truth on both sides. Pure, unmixed error does not live to plague us long. Error, when it continues to live, does so by virtue of a germ of truth contained. Great questions, therefore, continue to be argued *pro* and *con* from age to age, because each side is in a sense—i. e., from its own point of view—true, but wrong in excluding the other point of view ; and a true solution, a true rational philosophy, will always be found in a view which combines and reconciles the two partial, mutually excluding views, showing in what they are true and in what they are false—explaining their differences by transcending them. This is so universal and far-reaching a principle that I am sure I will be pardoned for illustrating it in the homeliest and tritest fashion. I will do so by means of the shield with the diverse sides, giving the story and construing it, however, in my own way. There is, apparently, no limit to the amount of rich marrow of truth that may be extracted from these dry bones of popular proverbs and fables by patient turning and gnawing.

We all remember, then, the famous dispute concerning the shield, with its sides of different colors, which we shall here call white and black. We all remember how, after vain attempts to discover the truth by dispute, it was agreed to try the scientific method of investigation. We all remember the surprising result. Both parties to the dispute were right and both were wrong. Each was right from his point of view, but wrong in excluding the other point of view. Each was right in what he asserted, and each wrong in what he denied. And the complete truth was the combination of the partial truths and the elimination of the partial errors. But we must not make the mistake of supposing that truth consists in *compromise.* There is an old adage that truth lies in the middle between antagonistic extremes. But it seems to us that this is the place of *safety,* not of truth. This is the favorite adage, therefore, of the timid man, the time-server, the fence-man, not the truth-seeker. Suppose there had been on the occasion mentioned above one of these fence-philosophers. He would have said : " These disputants are equally intelligent and equally valiant. One side says the shield is white, the other that it is black, now truth lies in the middle ; therefore, I conclude the shield is gray or neutral tint, or a sort of pepper-and-salt." Do we not see that he is the only man who has no truth in him ? No ; truth is no heterogeneous mixture of opposite extremes, but a stereoscopic combination of two surface views into one solid reality.

Now, the same is true of all vexed questions, and I

have given this trite fable again only to apply it to the case in hand.

There are three possible views concerning the origin of organic forms whether individual or specific. Two of these are opposite and mutually excluding; the third combining and reconciling. For example, take the individual. There are three theories concerning the origin of the individual. The first is that of the pious child who thinks that he was made very much as he himself makes his dirt-pies; the second is that of the street-gamin, or of Topsy, who says: "I was not made at all, I growed"; the third is that of most intelligent Christians—i. e., that we were made by a process of evolution. Observe that this latter combines and reconciles the other two, and is thus the more rational and philosophical. Now, there are also three exactly corresponding theories concerning the origin of species. The first is that of many pious persons and many intelligent clergymen, who say that species were made at once by the Divine hand *without natural process.* The second is that of the materialists, who say that species were not made at all, they were derived, "they growed." The third is that of the theistic evolutionists, who think that they were *created* by a process of evolution—who believe that making is not inconsistent with growing. The one asserts the divine agency, but denies natural process; the second asserts the natural process, but denies divine agency; the third asserts *divine agency by natural process.* Of the first two, observe, both are right and both wrong; each view is right in what it asserts, and wrong in what it denies—each is right from its own point of view, but wrong in excluding the other point of view. The third is the only true rational solution, for it includes, combines, and reconciles the other two; showing wherein each is right and wherein wrong. It is the combination of the two partial truths, and the elimination of the partial errors. But let us not fail to do perfect justice. The first two views of origin, whether of the individual or of the species, are indeed both partly wrong as well as partly right; but the view of the pious child and of the Christian contains by far the more essential truth. Of the two sides of the shield, theirs is at least the whiter and more beautiful.

But, alas! the great bar to a speedy settlement of this question and the adoption of a rational philosophy is not in the head but in the heart—is not in the reason but in pride of opinion, self-conceit, dogmatism. The rarest of all gifts is a truly tolerant, rational spirit. In all our gettings let us strive to get this, for *it* alone is true wisdom. But we must not imagine that all the dogmatism is on one side, and that the theological. Many seem to think that theology has a "*pre-emptive right*" to dogmatism. If so, then modern materialistic science has "*jumped the claim.*" Dogmatism has its roots deep-bedded in the human heart. It showed itself first in the domain of theology, because there was the seat of power. In modern times it has gone over to the side of science, because here now is the place of power and fashion. There are *two dogmatisms*, both equally opposed to the true rational spirit, viz., the old theological and the new scientific. The old clings fondly to old things, only because they are old ; the new grasps eagerly after new things, only because they are new. True wisdom and true philosophy, on the contrary, tries all things both old and new, and holds fast only to that which is good and true. The new dogmatism taunts the old for credulity and superstition ; the old reproaches the new for levity and skepticism. But true wisdom perceives that they are both equally credulous and equally skeptical. The old is credulous of old ideas and skeptical of new ; the new is skeptical of old ideas and credulous of new. Both deserve the unsparing rebuke of all right-minded men. The appropriate rebuke for the old dogmatism has been already put in the mouth of Job in the form of a bitter sneer : " No doubt ye are the people, and wisdom shall *die* with you." The appropriate rebuke for the new dogmatism, though not put into the mouth of any ancient prophet, ought to be uttered—I will undertake to utter it here. I would say to these modern materialists, " No doubt ye are the men, and wisdom and true philosophy were *born* with you."

Let it be observed that we are not here touching the general question of the personal agency of God in operating Nature. This we shall take up hereafter. All that we wish to insist on now is that the process and the law of evolution does not differ in its relation to materialism from all other processes and laws of Nature. If the sustentation of the universe by the law of gravitation does not disturb our belief in God as the sustainer of the universe, there is no reason why the origin of the universe by the law of evolution should disturb our faith in God as the creator of the universe. If the law of gravitation be regarded as the Divine mode of sustentation, there is no reason why we should not regard the law of evolution as the Divine process of creation. It is evident that if evolution be materialism, then is gravitation also materialism; then is every law of Nature and all science materialism. If there be any difference at all, it consists only in this : that, as already said, here is the *last* line of defense of the supporters of supernaturalism in the realm of Nature. But being the last line of defense—the last ditch—it is evident that a yielding here implies not a mere shifting of line, but a change of base ; not a readjustment of details only, but a *reconstruction of Christian theology.* This, I believe, is indeed necessary. There can be little doubt in the mind of the thoughtful observer that we are even now on the eve of the greatest change in traditional views that has taken place since the birth of Christianity. But let no one be greatly disturbed thereby. For as then, so now, change comes not to destroy but to fulfill all our dearest hopes and aspirations ; as then, so now, the germ of living truth has, in the course of ages, become so encrusted with meaningless traditions which stifle its growth, that it is necessary to break the shell to set it free ; as then, so now, it has become necessary to purge religious belief of dross in the form of trivialities and superstitions. This has ever been and ever will be the function of science. The essentials of religious faith it does not, it can not, touch, but it purifies and ennobles our conceptions of Deity, and thus elevates the whole plane of religious thought.

It will not, of course, be expected of me to give, even in briefest outline, a system of reconstructed Christian thought. Such an attempt would be wholly unbecoming. Time, very much time, and the co-operation of many minds, bringing contributions from many departments of thought, is necessary for this. In a word, it can only itself come by a gradual process of evolution. But from the point of view of science some very fundamental changes in traditional views are already plain. Of these the most fundamental and important are our ideas concerning God, Nature, and man in their relations to one another. These will form the subject of the next three chapters.

CHAPTER III.

THE RELATION OF GOD TO NATURE.

WE have already said that evolution does not differ essentially from other laws of Nature in its bearing on religious belief. It only reiterates and enforces with additional emphasis what Science, in all its departments, has been saying all along. The difficulties in the way of certain traditional views have pressed with ever increasing force upon the thoughtful mind ever since the birth of modern science. All along, an issue has been gathering, but put off from time to time by compromise, until now, at last, the issue is forced upon us and compromise is exhausted. The issue (let us look it squarely in the face) is : Either God is far more closely related with Nature, and operates it in a more direct way than we have recently been accustomed to think, or else (mark the alternative) Nature operates itself and needs no God at all. There is no middle ground tenable.

Let us trace rapidly the growth of this issue. The old idea and the most natural to the religious mind was the direct agency of God in every event and phenomenon of Nature. This view is nobly expressed in the noblest

literature in the world—in the Hebrew and Christian
Scriptures : " He looketh on the earth and it trembleth.
He toucheth the hills and they smoke." " He maketh
his sun to rise on the evil and on the good, and sendeth
his rain on the just and on the unjust." But now comes
Science and explains all these phenomena by natural laws
and resident forces, and we all accept her explanation.
Thus, one by one the phenomena of Nature are explained
by the operation of resident forces according to natural
laws, until the whole course of Nature, as we now know
it, has been, or will be, or conceivably may be, thus explained.

Thus has gradually grown up, without our confessing
it, a kind of scientific polytheism—one great Jehovah,
perhaps, but with many agents or sub-gods, each independent, efficient, and doing all the real work in his own
domain. The names of these, our gods, are gravity, light,
heat, electricity, magnetism, chemical affinity, etc., and
we are practically saying : " These be your gods, O
Israel, which brought you out of the land of Egyptian
darkness and ignorance. These be the only gods ye
need fear, and serve, and study the ways of."

What, then, is practically the notion which most
people seem to have of the relation of Deity to Nature ?
It is that of a great master-mechanic far away above us
and beyond our reach, who once upon a time, long ago,
and once for all, worked, created matter, endowed it with
necessary properties and powers, constructed at once out
of hand this wonderful cosmos with its numberless

wheels within wheels, endowed it with forces, put springs in it, wound it up, set it a-going, and then—*rested*. The thing has continued to go of itself ever since. He might have not only rested but *slept*, and the thing would have gone of itself. He might not only have slept but *died*, and still the thing would have continued to go of itself. But, no, I forget. He must not sleep or die, for the work is not absolutely perfect. There are some things too hard even for Him to do in this masterful, god-like way. There are some things which even He can not do except in a 'prentice-like, man-like way. The hand must be introduced from time to time to repair, to rectify, to improve, especially to introduce new parts, such as new organic forms.

Such was the state of the compromise until twenty-five years ago. Nature is sufficient of itself for its *course* and continuance, but not for *origins* of at least *some* new parts. Such was the state of the compromise until Darwin and the theory of evolution. But, now, even this poor privilege of occasional interference is taken away. Now, origins, as well as courses, are reduced to resident forces and natural law. Now, Nature is sufficient of itself, not only for sustentation, but also for creation. Thus, Science has seemed to push Him farther and farther away from us, until now, at last, if this view be true, evolution finishes the matter by pushing Him entirely out of the universe and dispensing with Him altogether. This, of course, is materialism. But this is no new view now brought forward for the first time by evo-

21

lution. On the contrary, evolution only finishes what science has been doing all along.

See, then, how the issue is forced. Either Nature is sufficient of itself and wants no God at all, or else this whole idea, the history of which we have been tracing, is radically false. We have here given by science either a demonstration of materialism or else a *reductio ad absurdum*. Which is it ? I do not hesitate a moment to say it is a reductio ad absurdum. And I believe that evolution has conferred an inestimable benefit on philosophy and on religion by forcing this issue and compelling us to take a more rational view.

What, then, is the alternative view ? It is the utter rejection with Berkeley and with Swedenborg of the independent existence of matter and the real efficient agency of natural forces. It is the frank return to the old idea of direct divine agency, but in a new, more rational, less anthropomorphic form. It is the bringing together and complete reconciliation of the two apparently antagonistic and mutually excluding views of *direct agency* and *natural law*. Such reconciliation we have already seen is the true test of a rational philosophy. It is the belief in a God not far away beyond our reach, who once long ago enacted laws and created forces which continue of themselves to run the machine we call Nature, but a God *immanent*, a God resident *in* Nature, at all times and in all places directing every event and determining every phenomena—a God in whom in the most literal sense not only we but all things have their being, in whom all things consist, through whom all things exist, and without whom there would be and could be nothing. According to this view the phenomena of Nature are naught else than objectified modes of divine thought, the forces of Nature naught else than different forms of one omnipresent divine energy or will, the laws of Nature naught else than the regular modes of operation of that divine will, invariable because He is unchangeable. According to this view the law of gravitation is naught else than the mode of operation of the divine energy in sustaining the cosmos—the divine method of sustentation ; the law of evolution naught else than the mode of operation of the same divine energy in originating and developing the cosmos—the divine method of creation ; and Science is the systematic knowledge of these divine thoughts and ways—a rational system of natural theology. In a word, according to this view, there is no real efficient force but spirit, and no real *independent* existence but God.

But some will object that this is pure *Idealism*. Yes, but far different from what usually goes under that name. The ideal philosophy as usually understood regards the external world as having no real objective existence outside of *ourselves*—as objectified mental states of the *observer*—as literally such stuff as dreams are made of—as a mere phantasmagoria of trooping shadows having no real existence but in the mind of the dreamer, and each dreamer makes his own world. Not so in the idealism above presented. According to *this* the external world is the objectified modes, not of the mind of the observer, but of the mind of God. According to this, the external world is not a mere unsubstantial figment or dream, but for *us* a very substantial objective reality surrounding us and conditioning us on every side.

Again, it will be objected that this is pure *Pantheism*. Again, we answer "yes." Call it so if you like, but far different from what goes under that name, far different from the pantheism which sublimates the personality of the Deity into all-pervading unconscious force, and thereby dissipates all our hopes of personal relation with him. Properly understood, we believe this view completely reconciles the two antagonistic and mutually excluding views of impersonal pantheism and anthropomorphic personalism, and is therefore more rational than either. The discussion of this most important point can only come up after the next chapter, because the argument for the personality of Deity is derived, not from without by the study of Nature, but from within in our own consciousness. We therefore put off its discussion for the present.

But, finally, some will object, "We can not live and work effectively under such a theory unless, indeed, we escape through pantheism." It may, alas! be true that this view brings us too near Him in our sense of spiritual nakedness and shortcoming. It may, indeed, be that we can not live and work in the continual realized presence of the Infinite. It may, indeed, be that we must still wear the veil of a practical materialism on our hearts and minds. It may, indeed, be that in our practical life and scientific work we must still continue to think of natural forces as efficient agents. But, if so, let us at least remember that this attitude of mind must be regarded only as our ordinary work-clothes—necessary work-clothes it may be of our outer lower life—to be put aside when we return *home* to our inner higher life, religious and philosophical.

CHAPTER IV.

THE RELATION OF MAN TO NATURE.

THERE are two widely distinct views concerning the relation of man to Nature : the one as old as the history of human thought, the other only now urged upon us by modern science. According to the one, man is the counterpart and equivalent of Nature. He alone has— in fact is—an immortal spirit, and therefore he belongs to a world of his own. According to the other, man is but a part, a very insignificant part of Nature, and connected in the closest way with all other parts, especially with the animal kingdom. He has no world of his own, nor even kingdom of his own : he belongs to the animal kingdom. In that kingdom he has no department of his own : he is a vertebrate. In the department of vertebrates he has no privileged class of his own : he is a mammal. In the class of mammals he has no titled order of his own : he is a primate, and shares his primacy with apes. It is doubtful if he may enjoy the privacy of a family of his own—the Hominidæ—for the structural differences between man and the anthropoid apes are probably not so great as between the sheep family and the deer family.

Now it is evident that these two are only views from different points, psychical and structural. From the psychical point of view it is simply impossible to exaggerate the wideness of the gap that separates man from even the highest animals. From this point of view man must be set over as an equivalent, not only to the whole animal kingdom, but to the whole of Nature besides. From the structural point of view, on the contrary, it is impossible to exaggerate the closeness of the connection. Man's body is identified with all Nature in its chemical constituents, with the body of all animals in its functions, with all vertebrates, especially mammals, in its structure. Bone for bone, muscle for muscle, ganglion for ganglion, almost nerve-fiber for nerve-fiber, his body corresponds with that of the higher animals. Whether he was derived from lower animals or not, certain it is that his structure even in the minutest details is precisely such as it would be if he were thus derived by successive slight modifications.

Now, of these two views, the latter has been in recent times enormously productive in increasing our knowledge. Anatomy has become truly scientific only through comparative anatomy ; physiology through comparative physiology ; embryology through comparative embryology. Sociology is fast following in the same line, and becoming scientific through comparative sociology. Is not the same true also of psychology ? Will not psychology become truly scientific only through comparative psychology, i. e., by the study of the spirit of man in relation to what corresponds to it in lower animals ? But this view and this method, when pushed to what seems to many their logical conclusion, end in identification of man with mere animals, of spirit with mere physical and chemical forces, immortality with mere conservation of energy, and thus leads to blank and universal materialism. Thus, while it increases our knowledge, it destroys our hopes. Is there any escape ? There is. The two extreme views given above are not irreconcilable. As already said, they are only views from different points, and therefore, although both true, are equally one-sided and partial, and a true and rational philosophy, in this as in all other cases of vexed questions, is found only in a higher view, which combines and reconciles these mutually excluding extremes. Can we find such a view ? I think we can.

Let us first, however, trace some of the stages of this scientific materialism. There are two main branches of the argument for materialism : one derived from *brain-physiology*, the other from *evolution*. As we wish to be perfectly fair, we will present and even press the argument in both these directions, although the latter alone bears directly on the subject in hand.

In recent times, physiology has made great and, to many, startling advances in the direction of connecting mental phenomena with brain-changes. Physiologists have established the correlation of vital with chemical and physical forces,* and probably in some sense, at

---

* See an article by the author on this subject, "American Journal of

least, of mental with vital forces. They have proved, in every act of perception, first a physical change in a nerve-terminal, then a propagated thrill along a nerve-fiber, and then a resulting change, physical or chemical, in the brain ; and in every act of volition, a change first in a brain-cell, then a return thrill along a nerve-fiber, and a resulting contraction of a muscle. Even the velocity of the transmission to and fro has been measured, and the time necessary to produce brain-changes estimated. They have also established the existence of physical and chemical changes in the brain corresponding to every change of mental state, and with great probability an exact quantitative relation between these changes of brain and the corresponding changes of mind. In the near future they may do more : they may localize all the different faculties and powers of the mind, each in its several place in the brain, and thus lay the foundations of a truly scientific phrenology. In the far-distant future we may possibly do much more. We may connect each kind of mental state with a different and distinctive kind of brain-change. We may find, for example, a right-handed rotation of atoms associated with *love,* and a left-handed rotation associated with *hate,* or a gentle sideways oscillation associated with *consciousness,* and a vertical pounding associated with *will.* Now, suppose all this, and even much more, be done in the way of associating, both in degree and in kind, mental changes

with brain-changes. What then? "Why," say the materialists, "we thereby identify *mind* with *matter*, mental forces with material forces. Thought, emotion, consciousness and will become products of the brain, in the
same sense as bile is a product of the liver, or urea a
product of the kidneys."

Such is, in brief, the argument. Now, the answer:
We may do all we have supposed and much more. We
may push our knowledge in this direction as far as the
boldest imagination can reach, and even then we are no
nearer the solution of this mystery of the relation of
brain-changes and mental changes than we are now.
Even then it would be impossible for us to conceive *how*
brain-changes produce mental changes or *vice versa*.
Physical changes in sense-organs, transmitted along
nerve-fibers, determine changes in brain-substance. So
much is intelligible. But now there appear—how it is
impossible to imagine—consciousness, thought, emotion,
etc.—phenomena of an entirely different order, belong-
ing to an entirely different world. So different, that it is
impossible to imagine the nature of the nexus between,
or to construe the one in terms of the other. Brain-cells
are agitated and thought appears: Aladdin's lamp is
rubbed, and the genie appears. There is just as much
intelligible causal relation between the two sets of phenomena in the one case as in the other.

Now, this mystery is not of the nature of those which
disappear under the light of knowledge. On the contrary, science only brings it out in sharper relief, and

emphasizes its absolute unsolvableness. Suppose an absolutely perfect knowledge, perfect in degree, but human in kind. Suppose an ideally perfect science—a science which has so completely subdued its domain, and reduced it to such perfect simplicity, that the whole cosmos may be expressed in a single mathematical formula —a formula which, worked out with plus signs, would give every phenomenon and event which shall ever occur in the future, and with minus signs every phenomenon and event which has ever occurred in the past. Surely, this is an ideally perfect science. Yet, even to such a science, the relation of brain-changes to mental states would be as great a mystery as now. It would even come out in stronger relief, because so many other apparent mysteries would disappear. Like the essential nature of matter or the ultimate cause of force, this relation lies evidently beyond the domain of science. It requires some other *kind* of knowledge than human to understand it.

But materialists insist so much on the identity of brain-physiology with psychology, that even at the risk of tediousness we will multiply illustrations in order, if possible, to make this point still clearer. Suppose, then, we exposed the brain of a living man in a state of intense activity. Suppose, further, that our senses were absolutely perfect, so that we could see every change, of whatever sort, taking place in the brain-substance. What would we see ? Obviously nothing but molecular changes, physical and chemical; for to the outside ob-

server there is absolutely nothing else there to see. But
the subject of this experiment sees nothing of all this.
His experiences are of a different order, viz., consciousness,
thought, emotions, etc. Viewed from the *outside*, there is
—there can be—nothing but motions; viewed from the
inside, nothing but thought, etc.—from the one side, only
physical phenomena; from the other side, only *psychical*
phenomena. Is it not plain that, from the very nature of
the case, it must ever be so? Certain vibrations of brain
molecules, certain oxidations with the formation of carbonic acid, water, and urea on the one side, and there
appear on the other sensations, consciousness, thoughts,
desires, volitions. There are, as it were, two sheets of
blotting-paper pasted together. The one is the brain,
the other the mind. Certain ink-scratches or blotches,
utterly meaningless on the one, soak through and appear
on the other as *intelligible writing*, but how we know
not, and can never hope to guess. But when the paste
dissolves, *shall the writing remain?* We shall see.

But some will object. There is nothing specially
strange and unique in all this, for the same mystery underlies the essential nature of all kinds of force and
matter, and therefore all phenomena. True enough,
but with this difference. Physical and chemical forces
and phenomena are indeed incomprehensible in their essential nature; but once accept their existence, and all
their different forms are mutually convertible, construa-
ble in terms of each other and all in terms of motion.
But it is impossible by any stretch of the imagination

to thus construe mental forces and mental phenomena. It may, indeed, be impossible to conceive *how came* the plane of material existence, but, standing on that plane, all phenomena fall into intelligible order. But there is another plane above this one, having no intelligible relation with it. We must climb up and stand on this before its phenomena fall into intelligible order. In a word, material forces and phenomena are, indeed, a mystery, but only of the *first order*. But mental and moral forces and phenomena are a mystery even from the standpoint of the other, and are therefore a mystery of the *second order*—a mystery within a mystery.

We repeat, then, with additional emphasis after this examination, that we can not imagine between physical and psychical phenomena a relation of cause and effect *in the same sense* in which we use these terms in physical science, although in some sense there is doubtless such a relation. If man were the only animal we had to deal with, there would be no standing ground left for materialism. But there is still another difficulty which sticks deeper. It is that suggested by the *law of evolution* and enforced by the *comparative method*.

Relation of Man to Animals.—Man, we say, is endowed with, *is*, in fact, an immortal spirit. What is spirit ? We know things only by their phenomena; what are the phenomena of spirit ? Consciousness, will, intelligence, memory, love, hate, fear, desire — surely these are some of them. But has not a dog or a monkey all these ? Pressed with this difficulty, some have in-

deed felt compelled to accord immortal spirit to higher
animals. But we can not stop here. If to these, then
also to all animals ; for we have here only a sliding scale
without break. Can we stop now and make it coextensive with sentiency ? No ; for the lowest animals and
lowest plants merge into each other so completely that
no one can draw the line between them with certainty.
We must extend it to plants also. Shall we stop here
and make immortal spirit coextensive with life ? We
can not ; for life-force is certainly correlated with, transmutable into, and derivable from, physical and chemical
forces. We must extend it into dead nature also.
Therefore, everything is immortal or none. Our boasted
immortality by continued extension becomes thinner and
thinner until it evaporates into thin air. It becomes
naught else than *conservation of energy*, and not, as we
had hoped, conservation of *self-conscious personality*.
This may be interesting as a scientific fact ; but of what
value to us personally is a continued existence of our
spiritual forces as heat, light, electricity, or any other
form of unconscious force ? Thus, then, if once we pass
the gap between man and the higher animals, there is no
possibility of a stopping-place anywhere.

Such is the difficulty presented by comparison in the
taxonomic series. Take now the *embryonic* series. Each
one of us, individually, was formed gradually by a process of evolution, from a microscopic spherule of protoplasm undistinguishable in structure from the lowest
forms of protozoal life. Now, in this gradual process of

evolution, where did immortal spirit come in? Was it in the germ-cell? Then why deny it to the protozoan? Was it at the quickening, or at the birth, or at the moment of first self-consciousness, or at some later period of capacity of abstract thought? Again, when it did come in, was it something superadded or did it grow out of something already existing in the embryo or the infant?

Or take the *evolution series* from protozoan to man. This we have already seen is similar in outline to the other two. Now, in the gradual evolution of the animal kingdom throughout all geological time, terminating in man, when did immortal spirit come in? Did it enter with life, or with sentient life, or somewhere in the ascending scale of animals, or with the advent of man? If with man, was it some new thing added at once out of hand, or did it grow out of something already existing in animals?

This last, we are persuaded, is the only tenable view— the only view that can effect that reconciliation between the two extreme, mutually excluding views now usually held, which, as already seen, is the true test of a rational philosophy. I believe that the spirit of man *was* developed out of the *anima* or conscious principle of animals, and that this, again, was developed out of the lower forms of life-force, and this in its turn out of the chemical and physical forces of Nature; and that at a certain stage in this gradual development, viz., with man, it *acquired* the property of immortality precisely as it now, in the individual history of each man at a certain stage, acquires the capacity of abstract thought. This is, in brief, the view which I wish to enforce. The reader must understand, however, that this is *my own view* only, a view for which I have earnestly contended for twenty years. It appeals, therefore, not to authority, but only to reason. I wish now to present it as briefly as possible.

First, then, I would draw attention to the fact that there is nothing wholly exceptional in such transformation with the sudden appearance of new powers and properties ; but, on the contrary, it is in accordance with many analogies in the lower forces, and therefore *a priori* not only credible but probable. For example, force and matter may be said to exist *now* on several distinct planes raised one above another. There is a sort of taxonomic scale of force and matter. These are, 1, the plane of elements ; 2, the plane of chemical compounds ; 3, the plane of vegetal life ; 4, the plane of animal life ; and 5, the plane of rational and, as we hope, immortal life. Each plane has its own appropriate force and distinctive phenomena. On the first operates physical forces, producing physical phenomena only — for the operation of chemical affinity immediately raises matter to the next plane. On the second plane operates, in addition to physical, also chemical forces, producing all those changes by action and reaction, the study of which constitutes the science of chemistry. On the third plane, in addition to the two preceding forces, with their characteristic phenomena, operates also life-force, producing the distinctive phenomena characteristic of living things. On the fourth plane, in addition to all lower forces and their phenomena, operates also a higher form of life-force characteristic of animals, producing the phenomena characteristic of sentient life, such as sensation, consciousness, and will. On the fifth plane, in addition to all the preceding forces and phenomena, we have also the forces and phenomena characteristic of rational and moral life.

Now, although there are doubtless great differences of level on each of these planes, yet there is a very distinct break between each. Although there are various degrees of the force characteristic of each, yet the difference between the characteristic forces is one of kind as well as of degree. Although energy by transmutation may take all these different forms, and thus does now circulate up and down through all these planes, yet the passage from one plane upward to another is not a gradual passage by sliding scale, but *at one bound*. When the necessary conditions are present, a new and higher form of force at once appears, *like a birth* into a higher sphere. For example, when hydrogen and oxygen are brought together under proper conditions, water is born—a new thing with new and wholly unexpected properties and powers, entirely different from those of its components. When CO_2, H_2O, and NH_3 are brought together under suitable conditions, viz., in the green leaves of plants, in the presence of sunlight, living protoplasm is then and there born, a something having entirely new and unexpected powers and

22

properties. It is no gradual process but sudden, like birth into a higher sphere.

Now, there is not the least doubt that the same is true of the order and manner of the *first appearance* of the natural forces in the phylogenic series. In the history of the evolution of the cosmos, the forces of Nature have appeared successively and suddenly when conditions became favorable. There was a time in the history of the earth when only physical forces existed, chemical affinity being held in abeyance by the intensity of the heat.* By gradual cooling, chemical affinity at a certain stage came into being—was born, a new form of force, with new and peculiar phenomena, though doubtless derived from the preceding. Ages upon ages passed away until the time was ripe and conditions were favorable, and life appeared —a new and higher form of force, producing a still more peculiar group of phenomena, but still, as I believe, derived from the preceding. Ages upon ages again passed away, during which this life-force took on higher and higher forms—in the highest foreshadowing and simulating reason itself—until finally, when the time was fully ripe and conditions were exceptionally favorable, spirit, self-conscious, self-determining, rational, and moral, appeared—a new and still higher form of force, but still, as I am persuaded, derived from the preceding.

Now, that these forces are really of derivative origin is proved by the fact that we see every step of this process

---

* All chemical compounds are dissociated by sufficient heat.

taking place daily under our very eyes. I pass over the conversion of physical into chemical force because this is admitted on all hands. I begin, therefore, with vital force. Sunlight falling on green leaves disappears as light and reappears as life—is consumed in doing the work of decomposing CO_2, H_2O, and NH_3, and the C, H, O, and N thus set free from previous combination unite to form living protoplasm.* Again, in the embryonic history of every animal we see the next change take place—i. e., the emergence of the psychic out of the vital. In the germ-cell, in the egg, and even in the early stages of the embryo, there is no distinctive animal life—i. e., no consciousness, nor volition, nor response of any kind to stimulus. At a certain stage distinctive animal or psychic life appears. We call it quickening. Materials for psychology are now present for the first time. In man alone, and that only some time after physical birth, we see the last change. The new-born child has animal life only. The emergence of self-consciousness—a change so wonderful that it may well be called the birth of spirit—takes place only at the age of two to three years. Now for the first time we have phenomena distinctive of humanity.

But some will ask, " How is this consistent with immortality? " In answer, let me again remind the reader

---

* The origin of vital from chemical force in the green leaves of plants can not be doubted; but this does not, of course, explain the mystery of the *first origin of life on the earth*, for one condition of the change *now* is the *contact of living matter*.

that with every new form of force, with every new birth
of the universal energy into a higher plane, there appear
new, unexpected, and, previous to experience, wholly unimaginable properties and powers. This last birth is of
course no exception. Why may not immortality be one
of these new properties? But this point is so important
that we must treat it more fully.

Remember, then, the view of the relation of God to
Nature, already explained. Remember that the forces of
Nature are naught else than different forms of the one
omnipresent Divine energy. Remember that, as just
shown, this Divine omnipresent energy has taken on successively higher and higher forms in the course of cosmic
time. Now this upward movement has been wholly by
increasing individuation, not only of matter, but also
of force. This universal Divine energy, in a generalized
condition, *unindividuated*, diffused, pervading all Nature, is what we call physical and chemical force. The
same energy in higher form, individuating matter, and
itself individuated, but only yet very imperfectly, is what
we call the life-force * of plants. The same energy, more

---

* I know it is the fashion to ridicule the use of the terms vitality,
vital force, as a remnant of an old superstition; and yet the same men
who do so use the terms gravity, electricity, chemical force, etc. Vital
force is indeed *correlated* with other forces of Nature, but is none the
less a distinct *form* of force, far more distinct than any other unless it
be the still higher form of psychical, and therefore it better deserves a
distinct name than any lower form. Each form of force gives rise to a
peculiar group of phenomena, and the study of these to a special department of science. Now, the group of phenomena called vital is more
peculiar, more different from other groups than these are from each

fully individuating matter and itself more fully individuated, but not completely, we call the *anima* of animals.
This anima, or animal soul, as time went on, was individuated more and more until it resembled and foreshad-
owed the spirit of man. Finally, still the same energy,
completely individuated as a separate entity and therefore
self-conscious, capable of separate existence and therefore
immortal, we call the spirit of man.

According to this view, the vital principle of plants
and the anima of animals are but different stages of the
development of spirit in the womb of Nature : *in man at
last it came to birth.* In plants and animals it was in
deep embryo sleep—in the latter, quickened, indeed, but
not viable—still unconscious of self, incapable of independent life, with physical, umbilical connection with
Nature ; but now at last in man, separated from Nature,
capable of independent life, born into a new and higher
plane of existence. Separated, but not wholly : Nature is
no longer *gestative* mother, but still *nursing* mother of
spirit. As the *organic embryo* at birth reaches independent *material* or *temporal* life, even so *spirit embryo* by
birth attains independent *spiritual* or *eternal* life.

Although birth is its truest correspondence and best
illustration, yet we may vary the illustration in many
ways :

---

other, and the science of physiology is a more distinct department than
either physics or chemistry, and therefore the form of force, which determines these phenomena, is more distinct and better entitled to a name
than any physical or chemical force.

1. Nature may be likened to a level water-surface, This represents unindividuated physical and chemical force. On this surface some individuating force pulls up a portion of the water into a *commencing* drop. This represents the condition of spirit in plants. Or by greater force the surface may be lifted higher into a nipple-like eminence *simulating a drop*, or even into an almost complete drop, with only a neck-like connection with the general surface. This represents the condition of spirits in the higher animals. In all these cases, even though the drop be nearly completed, if we remove the individuating or lifting force, the commencing drop is immediately drawn back by cohesion and refunded into the general watery surface. But, once complete the drop, and there is no longer any tendency to revert, even though the lifting force is removed. This represents the condition of spirit in man.

2. Or Nature may, again, be likened to a water-surface beneath which the anima of animals is deeply and tranquilly submerged, wholly unknowing of any higher, freer world above. In man spirit emerges above the surface into a higher world, looks down on Nature beneath him, around on other emerged spirits about him, and upward to the Father of all spirits above him. Emerged, but not wholly free—head above, but not yet foot-loose.

3. Or, again: As a planet must break away from physical, cohesive connection with the central sun (planet-birth) in order to enter into higher gravitative relations, which thenceforward determine all its movements in

beautiful harmony; as the embryo must break away from
physical umbilical connection with the mother in order
to enter into higher spiritual bonds of love, which
thenceforward determine all their mutual relations—even
so spirit must break away from physical and material
connection with the forces of Nature, which are but the
omnipresent Divine energy, in order thereby to enter
into higher relations of filial love to God and brotherly
love to man.

4. As the new-born child differs little in grade of
physical organization from the mature but unborn embryo, but at the moment of birth there is a sudden and
complete change, not so much in the grade of organization but in the whole plane of existence—a change absolutely necessary for further advance, for another cycle of
life; even so at the moment of the origin of man, howsoever this may have been accomplished, there may have
been no great change in the *grade* of *psychical* structure,
but yet a complete change in the *plane* of psychical life
—a change absolutely necessary for further advance, for
another cycle of evolution. In both cases there is a sudden entrance into a new world, the sudden appearance of
a new creature with entirely different capacities—a passing out of an old world, a waking up in a new and
higher. According to this view, man alone is a *child* of
God, capable of separate spirit-life—separate but not yet
wholly independent of Nature. As already said, Nature
is no longer gestative mother, but still nursing mother of
spirit—we are weaned only by death.

5. Or, again : As in passing up the *organic* scale, we find all grades of completeness of organic individuality, an increasing individuation of bodily form which completes itself as a perfect organic individual only in the higher animals, so, also, in passing up the *dynamic* scale, force or energy is individuated more and more until the process reaches completeness as a spirit-individual or dynamic individual—a person only in man. *Organic* individuality completes itself in animals. *Psychic* individuality only in man.

6. One more illustration and the last. The animal body may be likened to an exquisitely adjusted instrument of communication between two worlds—the material world without and the spiritual world within. The key-boards of this marvelous instrument are the nerve-terminals of the sense-organs in contact with the material world, and the brain-cells in touch with the spirit-world. External Nature plays on the one by sensation and determines changes in spirit. Spirit plays on the other by will and muscular contraction, and determines changes in external Nature. Now, in animals spirit is fast asleep or at most dreaming, or even perhaps somnambulistic, but at least unconscious of *self*, and acts only by stimulus—only responds in some sense automatically as sleepers do. In man spirit is wide awake and may respond automatically like animals, or may choose not to respond at all. Moreover, it acts freely in its own domain—the world of ideas—*without external stimulus ;* or of its own free-will may *initiate* changes in the ex-

ternal world.   With God all phenomena commence at
the *spirit-end*.   In animals all commence at the *matter-
end*, and by automatic response terminate in the same.
Man alone lives in both worlds, partakes of both natures,
and acts according to either method.

The more we reflect on this subject, the more we
shall be convinced that completed spirit individuality
explains, as nothing else can, all that is characteristic
of man.   It is this which constitutes person, or the
self-acting ego.   It is this which constitutes self-consciousness, free-will, and moral responsibility.   And out
of these, again, grows, the recognition of relations to
other moral beings and to God, and therefore ethics and
religion.   Out of these, also, grows the capacity of indefinite voluntary progress.   This also means separate life,
spirit-viability, or immortality.   Self-consciousness especially seems to me the simplest sign of separate entity or
spirit-individuality, and its appearance among psychical
phenomena *the very act of spirit-birth*.   We may imagine man to have emerged ever so gradually from
animals: in this gradual development the moment he
became conscious of self, the moment he turned his
thoughts inward in wonder upon himself and on the
mystery of his existence as separate from Nature, that
moment marks the birth of humanity out of animality.
All else characteristic of man followed as a necessary
consequence.   I am quite sure that, if any animal, say a
dog or a monkey, could be educated up to the point of
self-consciousness (which, however, I am sure is impos-

sible), that moment *he* (no longer *it*) would become a moral responsible being, and all else characteristic of moral beings would follow. At that moment would come personality, immortality, capacity of voluntary progress; and science, philosophy, religion, would quickly follow.

We have emphasized self-consciousness as the most fundamental sign of spirit-individuality; but a difference of exactly the same kind is found running through the whole gamut of human faculties as compared with corresponding faculties in animals. As animal consciousness is related to human self-consciousness, so exactly is animal will to human free-will, animal intelligence to human reason, animal sign-language to rational grammatical speech of man, constructive art of animals to true rational progressive art of man. In every one of these the resemblance is great, but the difference is immense, and not only in degree but also in kind. In every case it is like shadow and substance, promise and fulfillment, or, still better, it is like embryo and child. The change from one to the other is like to a birth into a higher sphere, the beginning of another cycle of evolution. We would like to follow this idea out in detail, but it would lead us beyond the scope of this work. Those who desire to do so we would refer to an article by the author on the " Psychical Relation of Man to Animals." *

But it will be objected that there are other births

---

* " Princeton Review " for May, 1884.

of energy from lower to higher condition; but such births do not insure continued existence in the higher condition. In the gradual evolution of energy described on page 316, when a portion rises from physical to chemical, from chemical to vital, or from vital to sentient, it does not remain ever after in the higher condition—there is no immortality on the higher plane. On the contrary, all these lower forms of energy are continually ascending and descending; transformation is downward as well as upward. Why should there be an exception in this last birth? In these successive upward metamorphoses of energy why should the last only be permanent? I answer: Because it reaches at last its final goal, viz., complete individuation, as free, self-acting spirit; it reaches again the spiritual plane from which it sprang, and becomes thereby a partaker of the Divine nature; because it comes at last into moral relations with the absolute—the Divine—and therefore above the plane of shifting changes. If the scale of energy be likened to a ladder with many rounds, reaching from the plane of matter to the plane of spirit, then so long as energy is on the ladder it ascends and descends; but, once it reaches the plane of free spirit, it is in a wholly new world in which eternal ascent is the law.

Perhaps I can best bring out the reasonableness of my view by comparing it with other possible alternative views.

There are three possible views as to the nature, the origin, and the destiny of the human spirit: (1.) That it

pre-existed always—uncreated, underived, eternal, both
ways—backward as well as forward. Therefore, as it
never began, so it will never end. It is *immortal of its
own right.* This is substantially the view of Plato, of
Leibnitz, and perhaps some other philosophers. (2.)
That it is derived from God *directly*—created at once
without natural process ; that at the moment of creation
of the first man Adam, and at some unknown time and
in some inscrutable way in the history of each individual, it was *injected into the body from the outside,* and
at the same time *endowed* with immortality. This, I
take it, is the orthodox view. (3.) That it was indeed
derived from God, but not directly; created indeed, but
only by natural process of evolution; that it indeed pre-
existed, but only as embryo in the womb of Nature,
slowly developing through all geological times, and finally
coming to birth as *living* soul in man. Thus it *attains*
immortality at a certain stage of development, viz., at
spirit-birth. This is the view I have striven to enforce.

I hold up these three views: Which is the more
rational ? The view of Plato—that of self-existent, uncreated, eternal spirit—I think few will entertain at this
time of the world's day. The usual orthodox view I
have shown is surrounded with insuperable difficulties ;
is wholly unscientific and irrational. What is there left
but the view presented above ? Plato is right in asserting
pre-existence, but wrong in denying creation. The usual
view is right in asserting creation, but wrong in denying
natural process. The view I have presented asserts pre-

existence in embryo and creation by natural process. It therefore combines and reconciles the two extreme views, and is more rational than either.

Some General Conclusions.—There are still two or three thoughts so closely connected with what we have already said that we can not pass them over:

1. We have seen that every mental state corresponds with a particular brain state, and every mental change with a brain change. We have, therefore, here, two series, physical and psychical, corresponding with each other, term for term. For every change in the one there is a corresponding change in the other, both in kind and amount. Now, is not this the test of the relation of cause and effect? It certainly is. Yes, there must be a causal relation here, even though we are not able to understand the nature of the causal nexus. But which is cause and which effect? If the view above presented be correct, then in animals *brain changes* are in all cases the *cause* of psychical phenomena. In man alone, and only in his higher activities, *psychic* changes precede and determines brain changes. In man alone brain changes are determined not only by external but by *internal* impressions. Man alone perceives not only objects—*material things*—but also relations and properties *abstracted* from the objects, i. e., *ideal things;* and, moreover, not only relations between objects, but also relations between relations or ideas. In man alone there is an inner world — microcosm—the *things* of which are *thoughts*, ideas, etc. This *self-acting power*

of spirit on *the things of itself*, instead of merely re-acting as played upon by external nature, is charac-teristic of man, and is a necessary result and a sign of severance, partial at least, of physical bond with Nature.

2. Again, I have used the term vital *principle*. I must justify it. I know full well that it is the fashion to ridicule the term as a remnant of an old superstition which regards vital force as a sort of supernatural entity unrelated to other forces of Nature. No one has striven more earnestly than myself to establish the correlation of vital with physical and chemical forces; * and yet, if the view above presented be true, there is a *kind* of justifica-tion even for the term vital *principle*—much more, vital *force*. There is a kind of reason and true insight in the personification of the forces of Nature, and especially of vital force. All forces, by progressive dynamic indi-viduation, are on the way toward entity or personality, but fully attain that condition only in man.

3. Again, to perceive relations and properties ab-stracted from material things, to form abstract or general ideas, to form not only *per*cepts but also *con*cepts, is ad-mitted to be a characteristic of man—a characteristic on which all our science and philosophy rest. From time immemorial the vexed question has been debated, " Have such abstract or general ideas any *real* existence, or are they mere *names* of figments of the mind ? " This is the famous question of *realism* and *nominalism*. Now,

---

* " Popular Science Monthly," December, 1873.

if our view be correct, then there is one most fundamental abstraction, viz., *self*, which is indeed a *reality*.
Self-consciousness is the direct recognition of the one
reality, spirit, of which all others are the sign and
shadow—the true reality which underlies and gives po-
tency to all abstractions or ideas. Do we not find in this
view, then, the foundation of a true realism, or rather a
complete reconciliation of realism and nominalism?

4. Thus, then, Nature, through the whole geological
history of the earth, was gestative mother of spirit,
which, after its long embryonic development, came to
birth and independent life and immortality in man. Is
there any conceivable meaning in Nature without this
consummation? All evolution has its beginning, its
course, its end. Without spirit-immortality this beautiful cosmos, which has been developing into increasing
beauty for so many millions of years, when its evolution
has run its course and all is over, would be precisely as if
it had never been—an idle dream, an idiot tale signifying
nothing. I repeat: Without spirit-immortality the cosmos has no meaning. Now mark: It is equally evident
that, *without this gestative method of creation of spirit*,
the whole geological history of the earth previous to man
would have no meaning. If man's spirit were made at
once out of hand, why all this elaborate preparation by
evolution of the organic kingdom? The whole evolution
of the cosmos through infinite time is a gestative process
for the birth of spirit—a divine method of the creation of
spirits.

Thus, again, man is born of Nature into a higher nature. He therefore alone is possessed of two natures—a lower, in common with animals, and a higher, peculiar to himself. The whole mission and life-work of man is the progressive and finally the complete dominance, both in the individual and in the race, of the higher over the lower. The whole meaning of sin is the humiliating bondage of the higher to the lower. As the *material* evolution of Nature found its goal, its completion, and its significance in man, so must man enter immediately upon a higher *spiritual* evolution to find its goal and completion and its significance in the ideal man—the Divine man. As spirit, unconscious in the womb of Nature, continued to develop by *necessary* law until it came to birth and independent life in man, so the new-born spirit of man, both in the individual and in the race, must ever strive by *freer* law to attain, through a newer birth, unto a higher life.

CHAPTER V.

In the two preceding chapters we have discussed the relation of God to Nature and of man to Nature. There is still another relation, if possible, of still more vital importance to us, viz., the *relation of God to man*. This, of course, introduces the question of revelation—a subject which I approach with some reluctance. I feel I am treading on holy ground, and must do so with shoes removed. If it be asked, How is evolution concerned with the subject of revelation ? I answer Evolution emphasizes and enforces the reign of law taught by all science, and makes it at last *universal*. Many conclude, therefore, that, if evolution be true, a belief in the possibility of any form of revelation is irrational. I do not think this follows, and I will give my reasons. I do so, however, very briefly, because we are not yet ready to formulate our views except in the most general way.

If man be indeed something more than a higher species of animal ; if man's spirit be indeed a spark of Divine energy individuated to the point of self-consciousness and recognition of his relation to God ; if spirit-

23

embryo, developing in the womb of Nature through all
geological time, came to birth and independent spirit-life
in man, and thus man alone is a *child of God* as well as a
product of Nature—if all this be true, then it is evident
that this wholly *new* relation requires also a wholly different mode of Divine operation. If God operates on
Nature only by regular processes, which we call *natural
laws*, then he *must* operate on spirit in a different and
a more direct way, and this we call *revelation*. If to the
student of Nature it is inconceivable that He should
operate on Nature except by natural laws (for this
is the name we give to His chosen mode of operation
there), then to the student of theology it is equally inconceivable, if our view of man be true, that He should not
operate on spirit in some more direct and higher way,
i. e., by revelation.

But some will ask, Is not this a palpable violation of
law ? I think not. All divine operations are, must be,
according to reason, i. e., according to law. The operation of the divine on the human spirit, i. e., revelation,
must therefore be according to law, but a higher law
than that which governs Nature, and, therefore, from *the
point of view of Nature*, supernatural. There is nothing
wholly unique in this. Life is a higher form of force
than the physical and chemical. Life-phenomena are
therefore super-physical, and if we confined the term
Nature to dead Nature they would be supernatural. So
the free, self-determined acts of spirit on spirit, even of
the spirit of man on the spirit of man, much more of the

Spirit of God on the spirit of man, may be according to law, and yet from the natural point of view be supernatural. It is true that, in the complex of phenomena, material and spiritual inextricably woven together, which go to make up human life, Science must ever strive to reduce as much as possible to material laws, for this is her domain, and she is bound to extend it; but, if our view of man be true, there will always remain a large residuum of phenomena—a whole world of phenomena—which will never yield, because clearly beyond her domain. Standing on the lower material plane, these phenomena are wholly super-material, and therefore incomprehensible from the material point of view. We must rise and stand on the higher plane before these also are reduced to law, but a higher law than that operating on the lower plane. If, therefore, science insists on banishing the supernatural from the realm of Nature, theology may reasonably insist on its necessity, *in this sense,* in the realm of morals and religion.

If, then, the direct influence of the Spirit of God on the spirit of man be what we call revelation, then there is evidently no other kind of revelation possible; and, furthermore, such revelation is given to all men in different degrees. It is given to all men as conscience; in greater measure to all great and good men as clearer perception of righteousness; in pre-eminent measure to Hebrew prophets and Christian apostles; but supremely and perfectly to Jesus alone. But there is, and in the nature of things there can be, *no test of truth but rea-*

son. We *must* fearlessly, but honestly and reverently, try all things, even revelations, by this test. We must not regard, as so many do, the spirit of man as the passive amanuensis of the Spirit of God. Revelations to man must of necessity partake of the imperfections of the medium through which it comes. As pure water from heaven, falling upon and filtering through earth, must gather impurities in its course differing in amount and kind according to the earth, even so the pure divine truth, filtering through man's mind, must take imperfections characteristic of the man and of the age. Such filtrate must be redistilled in the alembic of reason to separate the divine truth from the earthy impurities.

CHAPTER VI.

IT will be observed that the views presented in the
last three chapters are closely connected with one another,
and all conditioned on the "Relation of God to Nature,"
urged in Chapter III. Now it will doubtless be objected
to this view, especially as applied in Chapter IV on the
" Relation of Man to Nature," that it is naught else than
pure pantheism ; that it destroys completely the personality of Deity, and with it all our hopes of communion
with him, and all our aspirations of love and worship
toward him ; that, according to this view, God becomes
only the soul or animating principle of Nature, operating everywhere but unconsciously like the vital principle
of an organism ; that the whole cosmos becomes in fact a
great organism, developing under the operation of *resident* force according to *necessary* law, only that *we*
apotheosize this omnipresent force and call it God ; and
finally, that God is naught else than an abstraction,
created like other abstractions or general ideas wholly by
the human mind, and having no objective existence.

Furthermore, it will be said, that according to this view, this omnipresent unconscious energy individuates itself by necessary law of evolution more and more until it reaches, *for the first time in man*, self-consciousness and immortality, and thus that man himself is the only self-conscious immortal being in existence, and therefore the only being worthy of reverence and worship. Thus, this view leads to humanity-worship or rather to *self-worship*.

I feel the full force of this objection. I answer it as follows : I freely admit that, following up this scientific line of thought alone, we are carried strongly in the direction of pantheism. But there is nothing strange or exceptional in this. In all the deepest questions, single lines of thought inevitably carry us to extreme one-sided views. This seems to be the necessary result of the essentially two-fold nature of man, self-conscious spirit in a material body, the relation between which is, and must ever be, inscrutable. On this account there is and must be a fundamental antithesis in human philosophy, i.e., two lines of thought, the material and spiritual, which lead to two apparently irreconcilable views.* We have already seen that a rational philosophy, whenever we are able to reach such, is always found in a higher and more comprehensive view, which includes, combines, and reconciles two one-sided, partial, and mutually excluding views. But spirit and matter, or mind and brain, or God

---

* For a fuller statement of this antithesis, see an article by the author entitled " Evolution in Relation to Materialism," " Princeton Review," for March, 1881.

and Nature, is *the* fundamental antithesis which underlies and is the cause of all other lesser antitheses. This antithesis, therefore, is absolutely fundamental, and therefore forever irreconcilable. We must accept both sides, even though we can not clearly perceive the nature of their relation. We must be content with compromise where we can not effect complete reconciliation. We must frankly acknowledge that the antagonism is apparent only, and the result of the limitation of our faculties, and believe that, if we could only rise to a high enough point of view, like all other antitheses, this also would disappear in a rational philosophy.

Now, to apply these principles. No one, we admit, can form a clear conception of how immanence of Deity is consistent with personality, and yet we must accept both, because we are irresistibly led to each of these by different lines of thought. Science, following one line of thought, uncorrected by a wider philosophy, is naturally led toward the one extreme of pantheistic immanence; the devout worshiper, following the wants of his religious nature, is naturally led toward the other extreme of anthropomorphic personality. The only rational view is to accept both immanence and personality, even though we can not clearly reconcile them, i. e., immanence without pantheism, and personality without anthropomorphism. We have already seen in the third chapter, how following the scientific line of thought, we are logically driven to immanence. We wish now to show how, following another line of thought, we are as logically driven to personality. On this most difficult subject, however, all we are prepared to do is to throw out some brief suggestions, in the hope that they may be carried out more perfectly by some thoughtful reader ; scatter some seed-thoughts, in the hope that, falling haply on good soil, they may spring up and bear more fruit than I have been able to produce.

1. In the gradual individuation of the universal Divine energy described in Chapter IV, there must of course be a corresponding growth of a kind of independent self-activity which reaches completeness in man, and in fact constitutes what we call self-consciousness and free will. The exact nature of the relation of Deity or of the general forces of Nature to this gradually individuated portion, I do not undertake to define. And how this idea of partial self-activity comports with the absoluteness of Deity we can not clearly understand. But this fact need not specially disturb us here ; for this is only one branch of the wider question of the moral agency of man in relation to the absolute sovereignty of God, or the freedom of man in relation to necessary law in Nature.

2. **Personality behind Nature.**—We have already shown that, if the brain of a living, thinking man were exposed to the scrutiny of an outside observer with absolutely perfect senses, all that he would or could possibly see would be molecular motions, physical and chemical. But the subject himself, the thinking, self-conscious spirit, would experience and observe by introspection only consciousness, thought, emotion, etc. On the *outside*, only physical phenomena ; on the *inside* only psychical phenomena. Now,

must not the same be necessarily true of Nature also?
Viewed from the outside by the scientific observer, nothing is seen, nothing can be seen, there is nothing else to
be seen, but motions, material phenomena; but behind
these, on the other side, on the *inside,* must not there
be in this case also psychical phenomena, consciousness, thought, will; in a word, *personality?* * In the
only place where we do get behind physical phenomena,
viz., in the brain, we find psychical phenomena. Are
we not justified, then, in concluding that in all cases
the psychical lies behind the physical? The human
brain is a wonderful instrument, by means of which, in
some inscrutable way, viz., in our own experience, we do
get behind, on the other side, on the inside of some material phenomena, and in so far become partakers of the
Divine nature. But behind other phenomena of Nature
we may never hope to penetrate either by observation or
experience, but only in dim way by highest reason. Science, even in the case of the brain, can not pass from the
one kind of phenomena to the other. If she would study
the *inside* she must abandon the *outside*—she must abandon the microscope and take to introspection. If she
would study the phenomena of the higher platform, she
must leave the lower and climb up and stand on the
higher. If this be true of the brain where the two kinds
of phenomena are brought so close together, how much
more is it true of the phenomena of the cosmos. We

* Johnstone Stoney, "Nature," vol. xxxi, p. 422.

can never hope, either by observation or by experience, to pass beyond the veil. We must abandon the methods of science and reach it, if at all, in some other way. Not the clear-sighted but the pure-hearted shall see God in Nature.

Thus, then, we see that our own self-conscious personality behind brain phenomena compels us to accept consciousness, will, thought, personality behind Nature. Now I assert that, once get this abstract idea in the mind, and by a necessary law of thought it gradually expands without limit, and eventually reaches the form of infinite consciousness, will, thought, etc., and therefore of an infinite person. This law of indefinite expansion may be illustrated by the ideas of space and time. The animal, and, indeed, the infant, understands space and time only in their relation to itself, but has not yet abstracted these from their contents. This comes only with the birth of self-conscious personality. But, so soon as the abstract idea of space is acquired, by a necessary law of mental activity it expands without limit, and finally becomes the idea of infinite space. Similarly, so soon as the idea of time as abstracted from its contents is conceived, it inevitably expands without limit and grows into the idea of infinite time. So is it precisely with the idea of self-conscious personality. The animal or the very young child is indeed conscious of its body and of external objects in their mutual relations, but not of self, as abstracted from its contents. The animal never attains it, the child does. Now, so soon as this idea of self-conscious personality—of a spiritual entity underlying material phenomena—appears, by a necessary law of mental activity it expands without limit, and inevitably reaches the idea of an infinite self, an infinite person, God, behind the phenomena of Nature.

But some will object that this idea of infinite person-ality is inconceivable. True enough; but *the opposite is far more inconceivable.* The ideas of infinite space and infinite time are also inconceivable, yet we must accept them, because the idea of all space or all time being lim-ited is still more inconceivable; for if we think of space or time as limited, immediately there comes the ques-tion, "What is there beyond the limit?" There is therefore this wide difference between these two in-conceivables: the one is so only in the sense of tran-scending the power of our mind, but the other is un-thinkable, self-contradictory, absurd. So also is it with self-conscious personality. The idea of an infinite self, i. e., God, is indeed inconceivable, but only in the sense of transcending our power of comprehension; but the idea of the consciousness behind the cosmos as being limited or finite is more than inconceivable, it is unthinkable, self-contradictory, absurd; for immediately comes the ques-tion, "What is there beyond which limits it?" To the Greek mind Zeus was limited; therefore of necessity came also the idea of Fate, superior to and limiting Zeus himself. To them, therefore, Fate was the real God—the absolute.

3. Divine Personality.—1 have used the word person-ality as expressing the nature of God. But let me not be misunderstood. I well know we can not conceive clearly

of an infinite, unconditioned personality. Deeply considered, it seems nothing short of a contradiction in terms.
All I insist on is this: In our view of the nature of God,
the choice is not between personality and something
lower than personality, viz., an *unconscious force* operating Nature by *necessity*, as the materialists and panthe-
ists would have us believe; but between personality as we
know it in ourselves and something inconceivably *higher*
than personality. Language is so poor that we are
obliged to represent even *our* mental phenomena by physical images. How much more, then, the Divine nature
by its human image! Self-conscious personality is the
highest thing we know or can conceive. We offer him
the very best and truest we have when we call him a
Person; even though we know that this, our best, falls
far short of the infinite reality.

4. Cause in Nature.—We have thus far spoken only or
principally of self-consciousness, but the same precisely is
true of another essential attribute of personality, viz., *free-
will*. Every one admits causative force or forces operating in Nature. Science has shown that all the different
kinds of force are but different forms of one omnipresent
energy. Now, looking abroad on Nature from the outside, this omnipresent energy seems to modern science as
simply resident, inherent in matter itself, and therefore
as operating unconsciously and by necessity. But the
question occurs, " Whence did we get the idea of force,
energy, *causation?* " I answer unhesitatingly: We get
it not from without by observation of Nature, but from

within through consciousness ; not from the outside view, but from the inside view of phenomena. We can not conceive of phenomena without force, of effects without cause, because we are intensely conscious of being ourselves through our wills an active cause of external phenomena. If we were merely passive observers, not active causers of changes in the external world, then these external phenomena would seem to us merely to shift and change and succeed each in a certain order. We might note the order and determine the laws of sequence, and thus form a science ; but it would never enter into our minds to imagine any causal or dynamical nexus between them. In the mind of such passive observer, but not doer—thinker, but not worker—would be completely realized the only thorough-going and consistent materialistic philosophy, i. e., a philosophy in which, like Comte's, cause and force have no place—are in fact banished as a superstition from science. But the clear consciousness of essential energy, of causative force within, the certainty that we ourselves, through our wills and by the conscious exertion of force do determine changes in the external world, compels us to attribute all changes to causative force of some kind, and naturally enough, until the interference of science, to a personal will like our own. Thus by a necessary law we project our internal states into external Nature.

But see now the steps of evolution of this idea. At first, i. e., in the uncultured races, and also in childhood, external forces take the form of a personal will like our

own residing in *each object*, and controlling its phenomena as our wills control our bodily movements (fetich-
ism). Then, as culture advances, it takes next the form
of several personal wills controlling each the phenomena of a different department of Nature (polytheism).
Finally, in the highest stage of culture, it takes the
form of one personal will controlling the phenomena of
the whole cosmos (monotheism). To the religious but
unscientific mind in all these stages the personal will is
anthropomorphic. But we have already seen (Chapter
III) how anthropomorphism has been driven by science
from one department after another, until now at last
by evolution it is driven out of Nature entirely, and
to those following this line of thought alone, the phenomena of Nature are relegated to forces inherent in
matter, and operating by laws necessary and fatal; and
not only so, but material forces are made to invade
even the realm of consciousness, and reduce this also to
material laws. Thus the savage *e*jects his own conscious
personal will into every separate object of Nature; the
modern materialist *in*jects material forces into the realm
of consciousness. But, as already seen, a rational philosophy admits these two antithetic views, and strives to
combine and reconcile them. This reconciliation, as far
as it is possible for us, is found in a personal will immanent in Nature, and determining directly all its
phenomena.

Thus it is evident that the idea of a causal nexus
between successive phenomena is a primary conception,

and therefore ineradicable and certain. Even from the purest evolution point of view it must be true, for, if man's mind grew out of the forces of Nature, this idea must represent a fact in Nature. Also, analysis shows that all causative force originates in *will*. Lastly, culture and reason, by a necessary law of expansion, carry us upward to the conception of one infinite sustaining and creative will. Science may sometimes obscure but can not destroy this idea. Evolution, which was supposed by some to have destroyed it for ever, has only temporarily obscured it in the minds of the unreflecting, by the supposed identity of evolution with materialism. From this temporary eclipse it now emerges with still greater clearness and far greater nobleness. For, observe : All the effects known to us in Nature are finite ; therefore a personal will, which determines these *separately* by successive acts, as we do, must also be finite like ourselves. But a will, which by *one eternal* act ever-doing, never done, determines the evolution and the sustentation of an infinite cosmos, must itself be infinite. Thus only in the doctrine of universal evolution do we rise to a just conception of God as an infinite cause.

5. **Design in Nature.**—As the idea of *cause* and force is related to *will*, so precisely is the idea of *design* related to *thought*. We get this also, not from without, but from within. Adaptation of means to ends is in our experience the result of thought, and we can not conceive it to result otherwise. The effect of science can not be to de-

stroy this primary conception—which, indeed, like all primary conceptions, is ineradicable, and already more certain
than anything can be made by proof—but only to exalt and
purify our conceptions of the designer. For, observe : In
any case of *adaptive* structure, whether in the animal body
or in planetary relations, the evidence of design is not in
the materials, but in the *use* of the materials ; not in the
parts, but in the *adjustment* of the parts for a purpose.
Design, purpose, adjustment, *adaptation*, are not material things, but relations or intellectual things, and
therefore perceivable only by thought, and conceivable
only as the result of thought. It is simply impossible
to talk about such adaptive structures without using language which implies design. The very word *" adaptive "*
implies it. It is impossible even to think of such structures without implicitly assuming intelligence as the
cause. It makes no particle of difference *how* the material originated, or whether it ever originated at all ; it
matters not whether the adaptation was done at once out
of hand, or whether by slow process of modification ; it
matters not whether the adaptive modification was
brought about by a process of natural selection, or by
pressure of a physical environment ; whether without law
or according to law. The removal of the result from man-
like directness of separate action can not destroy the idea
of design, but only modify our conception of the Designer.
What science, and especially evolution, destroys, therefore, is not the idea of design, but only our low anthropomorphic notions of the mode of working of the Designer.

Precisely the same change takes place here under the influence of science as has taken place in all our notions concerning God. The uncultured savage sees a *separate* god in every object. As culture advances, his gods become fewer and nobler, until, in the most advanced states, man recognizes but one infinite God, the creator and sustainer of all. God is still in every phenomenon, but no longer as a separate God, but only as the separate manifestation of the One. Thus culture takes away our gods, but only to compel us to seek him in nobler forms until we reach the only true God. But, even after the conception of the one God is reached, how many seem to regard him as altogether such a one as ourselves ; but science shows us that his ways are not like our ways, nor his ends as our ends. Thus science, more than all other kinds of culture, simplifies while it infinitely ennobles and purifies our conceptions of Deity.

Again, the same change takes place in our sense of *mystery*. I suppose most people imagine that it is the special mission of science to destroy all mystery. Many seem to think that superstition, or even religion, is inseparably connected with ignorance and mystery, and all must disappear together before the light of science. But not so. There is only a gradual progressive change —an evolution in the form of mystery as well as in the form of religion. To the savage everything is a *separate* mystery. The function of science is, indeed, to destroy these separate mysteries, by explaining them ; but, in doing so, it only reduces them to fewer and

24

grander mysteries, and these again to still fewer and grander, until, in an ideally perfect science, all separate and partial mysteries are swallowed up in the one all-embracing infinite mystery—the mystery of existence. There is still mystery in each object, but no longer a separate mystery—only a separate manifestation of the one overwhelming mystery.

Or, again, and finally : The same change occurs in our ideas of *creation*. At first every object is a separate creation—a manufacture. With advancing science these separate, creative acts become fewer and nobler, until now, at last, in evolution, all are embraced and swallowed up in *one eternal* act of creation—a never-ceasing procession of the divine energy. Every object is still a creation, but not a separate creation—only a separate manifestation of the one continuous creative act.

Now, precisely the same change must take place in our conception of design in Nature. To the uncultured there is a distinct and separate design in every separate work of Nature. But, as science advances, all these distinct, separate, petty, man-like designs are merged into fewer and grander designs, until, finally, in evolution at last, we reach the conception of the one infinite, all-embracing design, stretching across infinite space, and continuing unchanged through infinite time, which includes and predetermines and absorbs every possible separate design. There is still design in everything, but no longer a separate design—only a separate manifestation of the one infinite design.

Thus, then, our own self-consciousness and will and thought give rise, necessarily, to the conception of an infinite self-consciousness, will, and thought—i. e., God. The necessity to believe in self-conscious spirit behind bodily phenomena compels us to believe also in an infinite self-conscious spirit behind cosmic phenomena. Looking at the operations of this ever-active spirit, whether in the one case or the other, *from the outside*, it looks like unconscious energy inherent in matter itself, and therefore like necessity, or fate. But, looked at from the inside *in the one case*, the brain, we perceive only self-conscious, free activity of spirit. Therefore, we are compelled to acknowledge in the other case, the cosmos, also, the same source of all activity, the same cause of all phenomena. We are compelled to acknowledge an infinite immanent Deity behind phenomena, but manifested to us on the outside as an all-pervasive energy. But some portion of this all-pervasive energy again individuates itself more and more, and therefore acquires more and more a kind of independent self-activity which reaches its completeness in man as self-consciousness and free-will. We said, "*a kind of* independent self-activity." How this comports with the absoluteness of God we can not understand, any more than we can understand how it comports with invariable law in Nature. We simply accept them both as primary truths, even though we can never hope to reconcile them completely, because we can not understand the exact nature of the relation of spirit to matter. We can not look at the outside and the inside at the same time. If we could understand the relation of psychical phenomena to brain-changes, then might we hope to understand far more perfectly than now the relation of God to Nature. But as in the one case, the brain, although we can not understand the *nature* of the relation, yet we are sure of the intimacy of the connection of the two series, psychical and physical, term for term ; so in the other case, the cosmos, although we can not understand the exact *nature*, we are sure of the intimacy of the connection, *term for term*—every material phenomenon and event with a corresponding psychical phenomenon as its cause.

CHAPTER VII.

THE doctrine of the Divine immanency carries with
it the solution of many vexed questions. In fact, in its
light these questions simply pass out of view as no longer
having any significance. Several of these questions have
been alluded to in an indirect way in the previous chapter and in Chapter III. We take them up distinctly here,
and show their relation to evolution.

Religious thought, like all else, is subject to a law of
evolution, and therefore passes through regular stages.
Of these stages, three are very distinct and even strongly
contrasted. They correspond in a general way to the
three stages of Comte, which he has misnamed the *theological*, the *metaphysical*, and the *positive*. We will illustrate by many examples.

I. *Conception of God.*

This, the most fundamental conception of all religion,
has passed from a gross anthropomorphism to a true
spiritual theism, and the change is largely due to science

and especially to the theory of evolution. There are three main stages in the history of this change : (1.) The first is a *low* anthropomorphism. God is altogether such a one as ourselves, but larger and stronger. His action on Nature, like our own, is *direct ;* his will is wholly man-like, capricious and without law. (2.) The second is still anthropomorphism, but of a nobler sort. God is not *altogether* like ourselves. He is man-like; yes, but also *king-like.* He is *not* present in Nature, but sits enthroned above Nature in solitary majesty. He acts on Nature, not directly but indirectly, through physical forces and natural laws. He is an absentee landlord governing his estate by means of appointed agents, which are the natural forces and laws established in the beginning. He interferes personally and by direct action only occasionally, to initiate something new or to rectify something going wrong. This idea culminated and found the clearest expression in the eighteenth century, and was the necessary result of the scientific ideas then prevalent, viz., ideas of pre-established *stability* of cosmic order and *fixedness* of organic types. God was the great *artificer*, the great *architect*, working, as it were, on foreign material and conditioned by its nature. He established all things as they are in the beginning, and they have continued so ever since.

This conception still lingers in the religious mind, and is in fact the prevailing one now. It is a great advance on the preceding, but, alas! it removes God beyond the reach of our love. He is the architect of worlds, the artificer of the eye, the sovereign ruler of the universe, but not our Father. We are his creatures, his subjects, but not his children.

(3.) The third and last stage in this development is true spiritual theism. God is immanent, resident in Nature. Nature is the house of many mansions in which he ever dwells. The forces of Nature are different forms of his energy acting directly at all times and in all places. The laws of Nature are the modes of operation of the omnipresent Divine energy, invariable because he is perfect. The objects of Nature are objectified, externalized —materialized states of Divine consciousness, or Divine thoughts objectified by the Divine will. In this view we return again to *direct* action, but in a nobler, a spiritual, Godlike form. He is again brought very near to every one of us and restored to our love, for in him we live and move and have our being. In him all things consist, by him all things exist. This view has been held by noble men in all times, especially by the early Greek fathers, but is now verified and well-nigh demonstrated by the theory of evolution. No other view is any longer tenable.

The idea of God is of course the most fundamental of all religious ideas, and a change in this carries with it many other changes. Some of these necessary outcomes, especially the nature, the origin, and the destiny of the human spirit, and its relation to the Divine spirit, I have already treated in previous chapters. But there are others which flow so directly and obviously that they may be presented in brief space.

II. *Question of First and Second Causes.*

Among the most obvious of these is the question of first and second causes. This distinction, I suppose, did not exist in early thought. As a popular view, it was mainly due to the physical science of the eighteenth century. It was a necessary corollary of the idea of God as the great architect sitting outside of Nature and acting on Nature as on foreign material. According to this view, God is the original and primary cause of all things; but he *delegates* his power to *secondary* forces, such as gravity, heat, electricity, etc., which are therefore the immediate causes of phenomena. I believe that most persons hold this view still. But it is now being displaced by the idea of God immanent or resident in Nature as already explained. This view is a complete *identification* of first and second causes. All causes are mere modes of the first cause. They seem to us secondary, necessary, and unconscious only because they act according to invariable law. But law itself is only the mode of operation of a perfect will. Thus we have the same three stages of evolution here also: (1.) First, all is first cause, direct, man-like, capricious, lawless. (2.) Then the first cause acts king-like, indirectly by many appointed agents subject to pre-enacted laws. These agents or secondary causes directly determine all natural phenomena. (3.) Lastly, come the complete combination and reconciliation of these two. All is by first cause and direct action, like the first. All is by invariable law like the second, the law being only the mode of operation of a perfect will.

III. *Question of General and Special Providence.*

So also providence, general and special, is only another phase of the same question and solved in the same way. At first all is *special* providence—the result of caprice or favoritism and without law. Then all or nearly all is general providence operating by invariable law; but from time to time the general law is broken through for special purposes when necessary. Is not this the prevailing view now? Lastly, these two must be combined and reconciled in a third. All is alike general and special: general—i. e., according to law; special—i. e., by direct action. There is no real distinction between the two. The distinction vanishes in the light of a higher view.

IV. *The Natural and the Supernatural.*

In precisely the same category falls the question of the natural and the supernatural. The same three stages are evident here also, and the same solution: 1. First all is supernatural and lawless, and Nature is viewed with stupid wonder and abject fear. 2. Then Nature is reduced to mechanical laws and made subject to man. Wonder and fear give place to indifference and even perhaps to contempt. We practically live without God in the world. It requires, now, *miracles* or a violent breaking through of law in order to startle us out of our stupidity and awaken in us a sense of the Divine presence. 3. But we must come lastly to a higher philosophy. We must recognize that all is natural and all is supernatural according as we view it, but none more than another.

All is natural—i. e., according to law; but all is super-natural—i. e., above Nature, as we usually regard Nature, for all is permeated with the immediate Divine presence. Wonder in the contemplation of Nature returns, or rather exalted reverence and rational worship are given in place of open-mouthed wonder and superstitious fear. Once clearly conceive the idea of God permeating Nature and determining directly all its phenomena according to law, and the distinction between the natural and the super-natural disappears from view, and with it disappears also the necessity of miracles as *we usually understand mira-cles.* In fact, the word as we usually understand it has no longer any meaning.

I must stop a moment to explain, lest I be misunder-stood; and to enforce, lest it be thought I speak lightly.

Miracle, in the sense of violation of law, is simply im-possible, because law is the expression of the essential nature and perfection of God. It is as impossible for God to perform a miracle in this sense as it is for him to lie, and for the same reason, viz., that it is contrary to his essential nature. In what sense, then, is a miracle possible? I answer, only as an occurrence or a phe-nomenon *according to a law higher than any we yet know.* If we define Nature as phenomena governed by physical and chemical laws and forces, then life becomes super-natural and miraculous—because higher than Nature as we define it. If we reduce the phenomena of life to law and include these also in our definition of Nature but limit it there, then the free, self-determined phenomena of reason become supernatural because above our defini-tion of Nature. There may well be still other and higher modes of Divine activity, the law of which we do not and may never understand. These are above our present defi-nition of Nature, and therefore to us supernatural or miraculous. But, even if miracles in the ordinary sense were possible, is it not evident that the ordinary processes of Nature are far more wonderful, more truly Godlike, than any such miracle?

V. *Question of Design in Nature.*

So, again, the question of design or purpose or mind in Nature is similarly solved. It has been said, it is con-tinually now being said, that evolution has destroyed for-ever the teleological view of Nature—i. e., the idea of design in Nature. Yes, if we mean the man-like, cabinet-making, watch-making design of Paley and older writers —a separate petty design for each separate object. It has indeed destroyed this, but only to replace it by a far nobler conception—a truly Godlike design, a design em-bracing all space and running through all time, including and absorbing all possible separate designs and prede-termining them by a universal law of evolution.

Or the same question may be put in another way as "Mind *vs.* Mechanics in Nature." In the evolution of thought on this subject at first all was *mind*, but lawless, capricious, like our own. Then one department after another of Nature was reduced to mechanical, physical, necessary law, until all have been or will be or conceivably may be thus reduced, and mind seems driven out of Nature entirely. The friends of religion in despair cry out for at least some small corner left for mind. Thus I find in recent numbers of an English scientific periodical, "Nature," a discussion concerning mind as *one* of the factors of evolution.* Is it not amusing, if it were not so sad?—God the Divine mind as *one of the factors* of evolution! The true solution is very simple. All is mind or none; so also all is mechanics or none. It *is all mind through mechanics.* It is all mechanics from the outside; it is all mind from the inside. To science all is mechanics; to theology all is mind. It is the duty of philosophy to reconcile these two opposites by the higher view that mechanics is but the mode of operation of the Divine mind. There is only one form of evolution, viz., human progress, in which mind—but the *human*, not the Divine mind—is *one* of the factors of evolution. But to think and speak thus of God in relation to Nature is to place him on the human plane. It is gross anthropomorphism.†

VI. *Question of the Mode of Creation.*

I might multiply examples almost without limit, of questions the solution of which depends on this one of the relation of God to Nature. I give one more—Creation.

---

* "Nature," vol. xxxiv, p. 385. 1886.

† So, again, see a book recently published ("Nature," vol. xliii, p. 460, 1891), entitled "Whence comes Man, from Nature or from God?" The answer is plain. From both—from God through Nature. Evolution is the method of creation.

The creation of the universe *at once*—in the beginning—out of nothing—and then *rest ever since.* This old anthropomorphic idea is now replaced by that of continuous creation — unhasting, unresting, by an eternal process of evolution. For if the universal law of gravitation is the Divine mode of sustentation of the universe, the no less universal law of evolution is the Divine process of creation.

CHAPTER VIII.

THE RELATION OF EVOLUTION TO THE IDEA OF THE CHRIST.

WHAT think ye of Christ? This is indeed in many ways a test-question, and we ought frankly to meet it. I have feared heretofore to touch this question. I now only throw out some brief suggestions— scatter some seed-thoughts. Does Evolution have anything to say on this also? I think it does. This I proceed to show:

As organic evolution reached its goal and completion in *man*, so human evolution must reach its goal and completion in the *ideal man*—i. e., the Christ. According to this view, the Christ is the ideal man, and therefore —(mark the necessary implication)—and therefore the Divine man. We are all as men (as contradistinguished from brutes)—we are *all*, I say, *sons of God ;* the Christ is the well-beloved Son. We are *all* in the image of God ; he is the express and *perfect image*. We are all partakers in various degrees of the Divine nature ; in him the Divine nature is completely realized. It is not necessary that the ideal man—the Christ—should be perfect in knowledge or in power; on the contrary, he must grow in wisdom and in stature, like other men; but he must be *perfect in character*. *Character is essential spirit*. All else, even knowledge, is only environment for its culture. In the dazzling light of modern science we are apt to forget this. Character is the *attitude* of the human spirit toward the Divine Spirit. If I should add anything to this definition, I would say it is spiritual *attitude* and spiritual *energy*. In the Christ this attitude must be wholly right; the harmony—the union with the Divine—must be perfect. This perfect union gives, of necessity, also fullness of spiritual energy.

Now, I wish to show that, although the Christ as thus defined must be human — yes, even more intensely human than any one of us—yet by the law of evolution we ought to expect him to differ from us in an inconceivable degree, and especially in a superhuman way. This I do by a series of illustrations.

We have said that the Christ is the ideal and therefore the Divine man—that he is the goal and completion of humanity. But in evolution a goal is not only a completion of one stage, but also the beginning of another and higher stage—on a higher plane of life with new and higher capacities and powers *unimaginable from any lower plane*. Let me illustrate :

1. As man is the ideal—the goal and completion of animal evolution, and yet is he also a birth into a higher plane of life—the spiritual ; so the Christ, the ideal man, may be only the goal and completion of human evolution,

and yet is he also a birth into a new and higher plane—
the Divine.

2. As the human spirit pre-existed in embryo in animals, slowly developing through all geological times, until it came to birth and immortality in man, so the Divine spirit is in embryo in man in various degrees of development, and comes to birth and completion of Divine life in the Christ.

3. As animals reached, finally, *conscious relations* with God in man, even so man reaches *union* with God in the Christ. As man, the ideal animal, is a union of the *animal* with the *spiritual;* so the Christ, the ideal of human evolution, is a union of the *human* and the *Divine.*

4. Finally: As with the appearance of man there were introduced new powers and properties unimaginable from the animal point of view, and therefore from that point of view seemingly supernatural—i. e., above their nature—so with the appearance of. the Christ we ought to expect new powers and properties unimaginable from the human point of view, and therefore to us seemingly supernatural—i. e., *above our nature.*

The Christ as defined above—i. e., as the *ideal man*— is undoubtedly a true object of rational worship. There are two and only two fundamental moral principles, viz., love to God and love to man. Both of these must be embodied in a rational worship. The one must be embodied in the worship of an Infinite Spirit—God; the other in the worship of the ideal man—the Christ.

But some one will object that, admitting all this, it is impossible that the goal, the ideal, should appear until the *end of the course* of evolution. To him I answer: This is indeed trķe of animal evolution, but not of human evolution. We have already seen (see p. 88 *et seq.*) that there is an essential difference in this regard between these two kinds of evolution. In addition to all the factors of organic evolution, in human progress there is a new and higher factor added, which immediately takes precedence of all others. This factor is *the conscious voluntary co-operation of the human spirit in the work of its own evolution.* The method of this new factor consists essentially in the formation, and especially in the *voluntary pursuit, of ideals.* In organic evolution *species* are transformed by the *environment.* In human evolution *character* is transformed by *its own ideal.* Organic evolution is by *necessary* law—human evolution is by voluntary effort, i. e., by *free* law. Organic evolution is *pushed* onward and upward from behind and below. Human evolution is *drawn* upward and forward from above and in front by the attractive force of ideals. Thus the ideal of organic evolution can not appear until the end ; while the attractive ideals of human evolution *must* come—whether only in the imagination or realized in the flesh—but must come somehow *in the course.* The most powerfully attractive ideal ever presented to the human mind, and, therefore, the most potent agent in the evolution of human character, is *the Christ.* This ideal must come—whether in the imagination or in the flesh I say not, but—must

25

come somehow *in the course* and not at the end. At the end the whole human race, drawn upward by this ideal, must reach the fullness of the stature of the Christ.

But it will be again objected that all ideals are relative and temporary; that we are in fact drawn onward and upward by many successive ideals, one beyond another, in the course. Ideals are but mile-stones which we put successively behind us while we press on to another; they are successive rounds of an infinite ladder which we put successively beneath us while we rise higher. This one also we shall eventually put behind us and pass on.

To this I have two answers: Admitted that in many ways such is the course of progress; but who has been able to reach this ideal and conceive a higher? When this one is reached and completely realized in our personal character, it will be time enough to propose another.

Again, it is true that in many ways we have advanced and are still advancing by the use of partial ideals; but this use of partial and relative ideals is itself in only a temporary stage of evolution. At a certain stage we catch glimpses of the *absolute* moral ideal. Then our gaze becomes fixed, and we are thenceforward drawn upward forever. The human race has already reached a point when the absolute ideal of character is attractive. This Divine ideal can never again be lost to humanity.

CHAPTER IX.

THE RELATION OF EVOLUTION TO THE PROBLEM OF EVIL.

THE problem of evil has tasked the power and baffled the skill of the greatest thinkers in every age. It would be folly in me to imagine that I can solve it. Its complete solution is probably impossible in the present state of science. Yet I can not doubt that on this, as on every important question relating to man, the theory of evolution will throw new and important light. All I can hope to do is to throw out some brief suggestions on the subject.

If evolution be true, and especially if man be indeed a product of evolution, then what we call evil is not a unique phenomenon confined to man, and the result of an accident, but must be a great fact pervading all nature, and a part of its very constitution. It must have existed in all time in different forms, and subject like all else to the law of evolution. Let us, then, trace rapidly some of the steps of this evolution.

1. *External Physical Evil in the Animal Kingdom.* —As already seen in previous chapters, the necessary condition of evolution of the organic kingdom is a *strug-*

gle for life—a conflict on every side, with a seemingly *inimical* environment and a survival of only the strongest, the swiftest, or the most cunning—in a word, the fittest. Now, suppose the course of organic evolution finished in the introduction of man, and from this vantage-ground we look back over the course and consider its result. Shall we call that evil which was the necessary condition of the progressive elevation which culminated so gloriously ? Evil doubtless it seemed to the individual, struggling animal, but is this worthy to be weighed in comparison with the evolution of the whole organic kingdom until it culminated in man ? Is it not rather a *good* in disguise ? I suppose human arrogance may be willing enough to admit it in *this* case, where animals only are sufferers.

2. *Physical Evil in Relation to Man.*—But organic evolution, completed in man, was immediately transferred to a higher plane, and continued as social evolution ; material evolution is transformed into psychical evolution ; unconscious evolution, according to *necessary* law, to conscious voluntary progress toward a recognized goal, and according to a *freer* law. But in this transformation the fundamental conditions of evolution do not change. Man also is surrounded on every side with what at first seems to him an *evil environment,* against which he must ever struggle or perish. Heat and cold, tempest and flood, volcanoes and earthquakes, savage beasts and still more savage men. What is the remedy—the only conceivable remedy ? Knowledge of

the laws of Nature, and thereby acquisition of power over Nature. But increasing knowledge and power are equivalent to progressive elevation in the scale of psychical being. This conflict with what seems an evil environment is, therefore, the necessary condition of such elevation. It is not too much to say that, without this condition, except for this necessity for struggle, man could never have emerged out of animality into humanity, or, having thus emerged, would never have risen above the lowest possible stage. Now suppose, again, this ideal to have been attained—suppose knowledge of physical laws and power over physical forces to be complete—suppose physical nature completely subdued, put beneath our feet, and subject to our will, and, from the high intellectual position thus attained, we look back over the whole ground and consider the result. Shall that be called evil which was obviously the necessary condition for attaining our then elevated position? Evil it doubtless seemed to the individuals who fell, and still seems to us who now suffer, by the way in the conflict; but is physical discomfort or even physical death of the individual to be weighed in comparison with the psychical elevation of the individual, and especially of the race? Evidently, then, physical evil even in the case of man is only *seeming* evil, but *real* good.

3. *Organic Evil—Disease.* — But there is a more dreadful form of evil than that which results from *external* physical nature—an evil far more subtle and difficult to understand, and therefore to conquer. I mean

internal organic evil—disease in its diversified forms and
with its attendant weakness and suffering, inscrutable
often in its causes, insidious in its approaches, contagious, infectious, spreading from house to house, carrying suffering and death in its course, and leaving sorrow
and desolation behind. Is there any remedy which can
transmute this evil into good ? There is. It is again
knowledge—knowledge of the laws, and power over the
forces, of *organic nature.* Is it not evident that complete
knowledge of the laws of health and the causes of disease would put this evil also under our feet ? Is it not
evident that a perfect knowledge of the laws of health,
and a perfect living according to these laws, would so
entirely subdue this evil that men would no longer die
except by natural decay or by accident ? Is it not evident, also, that the race will not attain this knowledge
unless it be forced upon us by the necessity of avoiding
the dread evil of disease ?

Now suppose, again, this ideal attained, suppose this
dread evil subdued by complete knowledge, and again
from our elevated intellectual position we look back over
the ground. Shall we call that evil which was the necessary condition of our intellectual elevation ? Evil,
doubtless, it seems to us individuals who have suffered
and are still suffering through our ignorance ; but is
such individual suffering or even individual death to
be weighed against the psychical elevation of the individual and evolution of the race ? Ought not the
individual to be willing to suffer thus much vicarious-

ly for the race? Is not this seeming evil also a *real* good?

May we not, then, confidently generalize? May we not say that all physical evil is good in its general effect —that every law of Nature is beneficent in its general operation, and, if sometimes evil in its specific operation, is so only through our ignorance? Partly by survival of the fittest, and partly by intelligence, man, like other animals, brings himself in accord with the laws of Nature, and thus appropriates the good and avoids the evil, and Nature becomes beneficent only. But, also unlike any other animal, man by rational knowledge makes the laws of Nature his servants, and uses them for his own purposes, thus increasing his power and elevating the plane of his life.

4. *Moral Evil.*—But there is still another form of evil, the most dreadful of all. This one may be called *the* evil, in some sense, *the only evil*. It is that of which all other forms are but the shadows cast backward and downward along the course of evolution and on lower stages of existence. This consummation of all evil is *sin — moral disease* — more dreadfully contagious and deadly than any organic disease. What shall we say now? Is there any rational explanation of this evil? Is there any possible reason or excuse for an all-wise, all-powerful Ruler afflicting man alone of all His creatures with this greatest of all evils? In all other cases, the individual and the race sacrifice themselves for a time *physically* for the sake of final spiritual elevation; but

this *is spiritual debasement*. In all other cases, there is a sacrifice in the *course* in order to attain the *goal*, but this is a missing of the goal itself. Is there any view which mitigates this evil, any philosophic alchemy which can transmute this evil into good ? Age after age the human mind has prostrated itself in helpless paralysis before this problem. Most thinkers have been content to say, "Thou hast ordered it so. Thou art good. It must be right." But many, and among them some of the best minds, have said, "Either God is not all-good, or else not all-wise, or else not all-powerful, or else there is no God at all." Does evolution shed any light on this dread problem ? I believe it does.

We have said that all other evils are but shadows of this one, cast backward and downward on earlier stages of evolution and lower forms of existence. But from the evolution point of view these earlier and lower forms of evil are rather to be regarded as *fore*shadowings of the reality to come. They are but earlier and lower stages of the evolution of the *same thing*—embryonic conditions of the now full-grown evil. If so, then the same law must apply here also, though, as we shall see, with a difference. Here, also, the individual as well as the race finds himself surrounded by what seems an evil environment, against which he must struggle. The spirit of man is inclosed and conditioned by a lower environment, which he must subdue or perish. Here, then, is again a deadly conflict : "a law in the members warring against the law of the spirit, and bringing it into captivity"; a law of selfism warring against the law of love, and bringing it into subjection; solicitations to debasement on the one hand, and solicitations to wrong others on the other. How shall it be overcome? What is the remedy? Again I answer, Knowledge of and conformity to the *laws of the moral world*. But, as in other cases, so in this: this knowledge of and conformity to law, which is the true goal of humanity, will not be attained unless it is forced upon us by necessity and in self-defense— i. e., by evil.

Now suppose, once more, this knowledge and conformity be complete, and the ideal of humanity be attained, and from this final and highest position we look back over the whole ground. Shall that be called evil which from the very nature of a moral being and the laws of evolution was obviously the necessary condition of attaining the goal? Shall we not from this final position call it a good in disguise? Evil, doubtless, it seems to us who suffer and stumble and mayhap fall by the way; but shall the mishap of the individual be weighed as an equivalent against the evolution of the race and the attainment of its goal?

Ah! there is the rub. It is all well enough to talk of sacrificing the *physical* individual to the race, but not so the *spiritual*. If we believe in the immortality of the human spirit, if we do indeed stand related to God in the manner explained in Chapter IV, then moral evil in the individual has an entirely peculiar and an eternal significance—then the individual human spirit has an infinite worth and can not be sacrificed to the race ; for the evolution of the race itself is only in order to the perfecting of individual human souls. What shall we say now ? I answer : The sacrifice is not necessary. There is in the realm of morals *alone* a way of escape—a saving element which redeems the individual without violating the law. Let me explain.

It will, I think, be admitted by all that *innocence* and *virtue* are two very different things. Innocence is a *pre-established,* virtue a *self-established,* harmony of spiritual activities. The course of human development, whether individual or racial, is from innocence through more or less discord and conflict to virtue. And virtue completed, regarded as a condition, is holiness, as an activity, is spiritual freedom. Not happiness nor innocence but virtue is the goal of humanity. Happiness will surely come in the train of virtue, but if we seek primarily happiness we miss both. Two things must be borne steadily in mind : virtue is the *goal of humanity ;* virtue can not be given, it must be *self-acquired.*

Now we have already seen that in all evil the remedy, which not only cures it but transmutes it into good, is knowledge of law and conformity of conduct thereto— a true science and a successful art—in a word, knowledge of the laws of God and obedience to these laws. In the physical world ignorance of these laws is necessarily fatal, but not so in the moral world. Ignorance here is not necessarily fatal though dangerous. By the very nature of a moral being, the essential thing is not knowledge but *character* or virtue—the *will* to know and the *effort* to obey. In the physical realm, knowledge is the goal; in the moral realm, knowledge is only in order to virtue. Therefore, in the case of the individual struggling with moral evil within and without, the victory is always in his power. If he fails, it is his own fault. His utmost effort in this field must be successful, because the result is not external, but internal and in the realm of moral freedom. The spirit of man is self-acting and in some sense, though not absolutely, self-existing, and can not be ruined except by its own act. In the moral world, where the goal is not knowledge but character, attainment must be in proportion to honest endeavor in the right spirit.

Evil, then, has its roots in the necessary law of evolution. It is a necessary condition of all progress, and pre-eminently so of moral progress. But some will ask, " Why could not man have been made a perfectly pure, innocent, happy being, unplagued by evil and incapable of sin ? " I answer : The thing is impossible even to omnipotence, because it is a contradiction in terms. Such a being would also be incapable of virtue, would not be a moral being at all, would not in fact be man. We can not even conceive of a moral being without freedom to choose. We can not even conceive of virtue without successful conflict with solicitations to debasement. But these solicitations are so strong and so often overcome us, that we are prone to regard the solicitations themselves as essential evil instead of our weak surrender to them.

All evolution, all progress, is from lower to higher plane. From a philosophic point of view, things are not good and evil, but only higher and lower. All things are good in their true places, each under each, and all must work together for the good of the ideal man. Each lower forms the basis and underlying condition of the higher ; each higher must subordinate the lower to its own higher uses, or else it fails of its true end. The physical world forms the basis and condition of the organic, yet the organism rises to a higher plane only by ceaseless conflict with and adaptation to the physical environment, which therefore seems in some sense evil. The organic world in its turn underlies and conditions and *nourishes* the rational moral world. As the senses are the necessary feeders of the intellect, so the appetites are the necessary feeders of the moral nature. Yes, even the lowest sensual appetites are the necessary basis and nourishers of our highest moral sentiments. And yet the struggle for mastery of the higher spiritual with the lower animal is often so severe that the latter seems to many as *essential evil* to be extirpated, instead of a useful *servant* to be controlled. This view is asceticism. Now the whole view of evil usually held is a kind of asceticism, and therefore, like asceticism, must be only a transition phase of human thought. All that we call evil both in the material and the spiritual world is good, so long as we hold it in subjection as servants to the spirit, and only becomes evil when we succumb. All evil consists in the dominance of the lower over the higher ; all good in the rational use of the lower by the higher. Asceticism may, indeed, be the best philosophy for some. If we can not subdue the lower nature, we must try to extirpate it, and thus at any cost set free the higher from humiliating bondage. If we can not practice the higher virtue of *temperance* in *all* things, we must even try the lower virtue of *total abstinence* in *some* things. If our right eye offends, we must not hesitate to pluck it out; but let us not imagine that one eye is better than two —let us clearly understand that thereby our spiritual nature is sadly maimed, and therefore that the highest virtue, which is spiritual beauty and strength, can not thus be attained. True virtue consists, not in the extirpation of the lower, but in its subjection to the higher. The stronger the lower is, the better, *if only* it be held in subjection. For the higher is nourished and strengthened by its connection with the more robust lower, and the lower is purified, refined, and glorified by its connection with the diviner higher, and by this mutual action the whole plane of being is elevated. It is only by action and reaction of all parts of our complex nature that true virtue is attained.

INDEX.

Acceleration, law of, 178.

African fauna explained, 204.

Agassiz, his greatest result, 29, 43; relation to evolution, 32, 37, 43; relation to Darwin, 46; compared with Kepler, 47.

Ages of geological history, 16.

Alpine species explained, 215.

Amphibians, development of, 150.

Analogy and homology, 99.

Anima of animals, 313, 317.

Animal architecture, styles of, 209.

Animal kingdom, primary divisions of, 107.

Animals, relation of man to, 311; spirit embryonic in, 311.

Antiquity of man, religion and, 282; of the earth, religion and, 281.

Aortic arches, proofs of evolution from, 151.

Arthropods, 132.

Artificial production of varieties, 222.

Australia, fauna and flora of, explained, 200; when isolated, 202.

Barriers limit faunal and floral regions, 188.

Beauty, origin of, 269.

Birds' tails, changes of, 274.

Brain, vertebrate, proofs of evolution from, 162; vertebrate, changes of, in phylogenic series, 168; relation to mind, 327, 338.

Brain-physiology as a basis for materialism, 306.

Branching tree illustrates evolution, 13–15, 18, 110, 250.

Brooks, W. K., on the cause of variations, 262.

Californian coast-islands, fauna and flora of, 211.

Causation, idea of, from within, 342.

Cause, first and second, 354.

Cells, somatic and germ, 93.

Centers of creation, specific, 194.

Cephalization, 171.

Chambers, his views on evolution, 34.

Changes slow at present, 266.

Christ, the, 359; relation of evolution to, 359; as an agent in human progress, 363.

Close-breeding, effects of, 236, 243.

Coast-islands of California, fauna and flora of, 211.

Comparison, method of, 41.

Conflict between religion and science, 280.

Continental faunas and floras, 188.

Continental island life, 208.

Continuity, law of, 53; law of, applied to inorganic forms, 54; to organic forms, 56.

Cope's law of acceleration, 178.

Creation, special, 30, 69; specific centers of, 194; changes in our notions of, 348; question of mode, 358.

Cross-breeding, law of, 236.

Cross-fertility of artificial varieties, 232.

Cross-sterility, 77, 234.

Cyclical movement, law of, 16, 22.

Darwin, relation to Agassiz, 46; compared with Newton, 48; factors of evolution discovered by, 74; objections to his theory of evolution, 76.

Derivation, origin of inorganic forms by, 54; origin of organic forms by, 56.

Design, idea of, from within, 345; argument from, not destroyed by evolution, 346; changes in our ideas of, 348; in Nature, question of, 357.

Differentiation, law of, 11, 19; law of, in embryonic development, 19; law of, illustrated, 144; of the animal kingdom illustrated, 176.

Disease, necessity of, 367.

Divine energy, forms of, 318.

Divisions of the animal kingdom, 117.

Dogmatism, theological and scientific, 293.

Domestication, changes produced by, 222.

Egg, development of, 3, 19.

Egyptian species unchanged in three thousand years, 265.

Embryology, proofs of evolution from, 148.

Environment, physical, 73.

Evil, problem of, relation of evolution to, 365; physical, necessity of, 366; a condition of progress, 366, 373; organic, necessity of, 367; moral, necessity of, 369.

Evolution, what is, 3, 8; scope of, 3; type of, 3, 8; examples of, 5; popularly limited to the organic kingdom, 7; progressive change in, 9; laws of, 11; illustrated by branching tree, 13–15, 18, 90, 250; misconception of, 14; produced by resident forces, 27; germs of the idea, 32; relation of Agassiz to, 32, 37, 43; Lamarck's views on, 33; Chambers's views on, 34; obstacle to, removed, 35; conflicting with religion imaginary, 45; how related to gravitation, 49; general evidences of, 53; artificial, 60; observed, 62; certainty of, 65; special proofs of, 67; factors of, 73, 81; human contrasted with organic, 88; monotypal and polytypal, 85; proofs of, from the vertebrate skeleton, 111; from the articulate skeleton, 132; from embryology, 148; from development of amphibians, 150; from aortic arches, 151; from vertebrate brain, 162; from rudimentary organs, 179; from geographical distribution of organisms, 183; explains geographical diversity, 195; objection to this view, 217; answer, 219; proofs of, from artificial modifications, 222; factors of, operative in domestication, 228; paroxysmal, 257; material, nearly completed, 267; thoroughly established, 275; relation to religion, 276, 282; relation to materialism, 284; necessitates great change in religious thought, 295; of forces, 315; relation to revelation, 331; pantheistic objection answered, 335; relation to problem of evil, 365.

Experimental method largely fails on plane of life, 40.

Factors of evolution, 73; their grades and order of introduction, 81; Lamarckian, 81; selection, 82-85; Darwinian, 83; rational, 86.

Faculties, evolution of, 23.

Faunas and floras, geographical, 183; continental, 188; marine, 192; special cases of distinct, 192; of Australia, 200; of Africa, 204; of Madagascar, 205; of continental islands, 208; of the coast-islands of California, 211;

26

of oceanic islands, 213; of lofty mountains, 215.

Fish-tails, changes of, in development, 172; in evolution, 174.

Fishes, age of, 17.

Floras and faunas, geographical, 183.

Force, vital, correlation of, 36; planes of, 314; evolution of, 315; idea of, from within, 342.

Forces, resident, evolution by, 27; of Nature are forms of Divine energy, 317; different planes of, 314.

Fore-limbs, vertebrate, homologies of, 113.

Generation, spontaneous, 15.

Geographical faunas and floras, 183; diversity, theory of, 193; diversity explained by evolution, 195; present diversity determined by Glacial epoch, 198; objection to this view, 217; answer, 219.

Geological record, imperfection of, 252.

Glacial epoch determined distribution of species, 195, 198, 215; changes during, in America, 198; in Europe, 199.

God, relation of, to Nature, 297; immanence of, in Nature, 300; relation of, to man, 326; personality of, 332; necessary belief in, 344; different forms of conception, 351.

Good and the true, relation of, 277.

Grasshopper, external anatomy of, 143.

Gravitation, relation of, to evolution, 49 ; and religion, 281.

Gyroscope, 288.

Heliocentric theory and religion, 280.

Hind-limbs, vertebrate, homologies of, 121.

Horse, genesis of, 126.

Homologies of vertebrate skeleton, 111 ; of vertebrate fore-limbs, 113 ; of vertebrate hind-limbs, 121 ; of articulate skeleton, 132.

Homology and analogy, 99 ; only within primary divisions, 108.

Hyatt, A., on Planorbis, 254.

Ideal, relative and absolute, 364.

Idealism, true and false, 301.

Immortality in accord with law, 316.

Individuality, organic, 325 ; spiritual, 325.

Innocence and virtue compared, 372.

Inorganic forms, law of continuity applied to, 54.

Intermediate forms between artificial varieties, 232.

Islands, continental and oceanic, 207.

Kepler compared with Agassiz, 47

Lamarck, evolutionary views of, 33, 74.

Law of differentiation, 11, 19; of progress of the whole, 13, 22 ; of cyclical movement, 16, 22; of continuity, 53 ; of continuity ap-

plied to inorganic forms, 54; to organic forms, 56 ; of differentiation illustrated, 144 ; of acceleration, 178; of cross-breeding, 218, 236.

Laws of evolution, 11, 19.

Lepidosiren, 101.

Life, nature of, 35 ; imperfectly subject to experiment, 40; relation of, to philosophy, 277.

Limbs, vertebrate, homology of, 113.

Links, connecting, 12, 57, 145 ; connecting, elimination of, 248 ; connecting, usually absent from geological faunas, 251.

Liquidambar, 218, 220.

Lobster, external anatomy of, 136.

Lungs, formation of, 100.

Madagascan fauna explained, 205.

Mammals, age of, 17.

Man, age of, 18; relation of, to Nature, 304 ; relation of, to animals, 311 ; spirit of, in relation to the forces of Nature, 313, 316 ; relation of God to, 331.

Marsupials, 201.

Materialism, relation of, to evolution, 284 ; basis for, in brain-physiology, 306 ; basis for, in evolution, 311

Methods, scientific, 38.

Migration favors diversification, 77.

Mind, relation of, to brain, 327, 338; versus mechanics in Nature, 340.

Miracles, question of, 356.

Mollusks, age of, 16.

Monotremes, 201.

Mystery, changes in our sense of, 347.

Nature, relation of God to, 297; immanence of God in, 300; relation of man to, 304 ; has no meaning without spirit, 329 ; mind *versus* mechanics in, 340.
Natural and supernatural, 355.
Neo-Darwinism, 93 ; relation of, to human progress, 97.
Newton compared with Darwin, 48.
Nominalism and realism reconciled, 329.

Obstacle to evolution removed, 35.
Oceanic island life, 213.
Ontogenic series, 9, 40.
Organic forms, views of origin of, 29, 68, 72, 292 ; law of continuity applied to, 56.
Organs, incipient, 270.
Origin of varieties unexplained, 270.

Pantheism, true and false, 302, 335.
Paroxysmal evolution, 257.
Personality behind Nature, 338.
Personality of God, 337, 341.
Philosophy and life, relations of, 277.
Phylogenic series, 10, 41.
Planorbis of Steinheim, 254.
Primal animals, 145.
Progress of the whole, law of, 13, 22.
Progressive change in evolution, 9.
Providence, question of general and special, 355.

Ranges of organic forms, 186.
Realism and nominalism reconciled, 329.
Record, geological, imperfection of, 252.
Religion, so-called conflict of, with evolution, 45, 280.
Religious thought to be reconstructed, 295.
Reproduction, methods of, 237.
Reptiles, age of, 17.
Revelation, relation of evolution to, 331; not inconsistent with the laws of Nature, 332; nature of, 333.
Reversion of artificial forms, 229.
Romanes, G. J., his idea of physiological selection, 76, 84 ; the idea applied, 245.
Rudimentary organs, proofs of evolution from, 179 ; organs in man, 181.

Selection, sexual, 74, 85 ; natural, 74, 79, 83 ; physiological, 75, 79, 84; natural, compared with artificial, 225 ; physiological, applied, 245.
Self-consciousness the sign of spirit-individuality, 325.
Sequoia, 219, 220.
Sexes, characters of, compared, 262.
Shrimp, external anatomy of, 134.
Sin a condition of moral evolution, 350.
Skeleton, vertebrate, homologies of, 111; articulate, homologies of, 132; articulate, general structure of, 134.

Society, progress of, 25.

Space and time the two fundamental conditions of material existence, 48.

Species, natural, more permanent than artificial varieties, 229; more distinct, 232; cross-sterile, 232.

Spirit embryonic in animals, 311; of man related to *anima* of animals, 313; to forces of Nature, 313, 316; origin of illustrated, 320–322; Plato's view, 326; orthodox view, 326; no meaning in Nature without, 329.

Steinheim, Planorbis of, 254.

Supernatural and the natural, 355.

Taxonomic series, 9, 40.

Temperature-regions, 184.

Tread, plantigrade and digitigrade, 123.

True and the good, relation of, 277.

Truth tested by effect on life, 277; not compromise, 291.

Types, generalized, 13.

Use and disuse of organs, 73.

Useless structures, how produced, 76.

Variation depends on sexual reproduction, 238; caused by unfavorable conditions, 264.

Varieties, artificial production of, 222, 235; artificial production of, illustrated, 224; natural and artificial, compared, 228; origin of, unexplained, 270.

"Vestiges of Creation," 34.

Virtue and innocence compared, 372.

Vital principle, 328.

Voluntary social progress, 26.

Weismann's views, 93.

Whales, rudimentary organs of, 180.

THE END.